The Complete Guide to
Building Your Own
Greenhouse

Everything You Need to Know Explained Simply

By Craig Baird

THE COMPLETE GUIDE TO BUILDING YOUR OWN GREEN-HOUSE: EVERYTHING YOU NEED TO KNOW EXPLAINED SIMPLY

Library of Congress Cataloging-in-Publication Data

Baird, Craig W., 1980-
  The complete guide to building your own greenhouse : everything you need to know explained simply / by: Craig Baird.
     p. cm.
  Includes bibliographical references and index.
  ISBN-13: 978-1-60138-368-6 (alk. paper)
  ISBN-10: 1-60138-368-1 (alk. paper)
  1. Greenhouses--Design and construction. I. Title.
  SB416.B35 2011
  690'.8924--dc22
                                    2010043861

EDITOR: Sylvia Maye • EDITORIAL ASSISTANTS: Brittany Miller & Leslie Bream
INTERIOR LAYOUT: Antoinette D'Amore • addesign@videotron.ca
COVER DESIGN: Meg Buchner • meg@megbuchner.com
BACK COVER DESIGN: Jackie Miller • millerjackiej@gmail.com
PROOFING : Katy Doll • kdoll413@gmail.com

Printed in the United States

Printed on Recycled Paper

We recently lost our beloved pet "Bear," who was not only our best and dearest friend but also the "Vice President of Sunshine" here at Atlantic Publishing. He did not receive a salary but worked tirelessly 24 hours a day to please his parents. Bear was a rescue dog that turned around and showered myself, my wife, Sherri, his grandparents Jean, Bob, and Nancy, and every person and animal he met (maybe not rabbits) with friendship and love. He made a lot of people smile every day.

We wanted you to know that a portion of the profits of this book will be donated to The Humane Society of the United States. *–Douglas & Sherri Brown*

The human-animal bond is as old as human history. We cherish our animal companions for their unconditional affection and acceptance. We feel a thrill when we glimpse wild creatures in their natural habitat or in our own backyard.

Unfortunately, the human-animal bond has at times been weakened. Humans have exploited some animal species to the point of extinction.

The Humane Society of the United States makes a difference in the lives of animals here at home and worldwide. The HSUS is dedicated to creating a world where our relationship with animals is guided by compassion. We seek a truly humane society in which animals are respected for their intrinsic value, and where the human-animal bond is strong.

Want to help animals? We have plenty of suggestions. Adopt a pet from a local shelter, join The Humane Society and be a part of our work to help companion animals and wildlife. You will be funding our educational, legislative, investigative and outreach projects in the U.S. and across the globe.

Or perhaps you'd like to make a memorial donation in honor of a pet, friend or relative? You can through our Kindred Spirits program. And if you'd like to contribute in a more structured way, our Planned Giving Office has suggestions about estate planning, annuities, and even gifts of stock that avoid capital gains taxes.

Maybe you have land that you would like to preserve as a lasting habitat for wildlife. Our Wildlife Land Trust can help you. Perhaps the land you want to share is a backyard—that's enough. Our Urban Wildlife Sanctuary Program will show you how to create a habitat for your wild neighbors.

So you see, it's easy to help animals. And The HSUS is here to help.

THE HUMANE SOCIETY
OF THE UNITED STATES.

2100 L Street NW • Washington, DC 20037 • 202-452-1100
www.hsus.org

Trademark Disclaimer

Table of Contents

# Introduction

> *"Who loves a garden loves a greenhouse too."*
>
> — William Cowper, English Poet

L egend has it that as a child, America's first president, George Washington, chopped down a cherry tree. Honest to a fault, he confessed. "I can't tell a lie. I did cut it with my hatchet."

Regardless of this tale's veracity, Washington more than made up for it later. It is well known that he was a dedicated and proud gardener. In fact, on November 23, 1794, he wrote to William Pearce, and said, "I have no objection to any sober or orderly person's gratifying their curiosity in viewing the buildings, Gardens, & ca. about Mount Vernon."

He devoted large areas of his Mount Vernon home to planting, and when he was able to be at home, he and his gardeners tested their skills by growing plants that were exotic to Virginia in a greenhouse, which provided a winter refuge for tropical and semi-tropical plants, such as oranges, lemons, limes, coffee and aloe. A fire was maintained to keep the greenhouse warm.

In June of 1787, around the time Washington was inaugurated as America's first president, the British ship, HMS Bounty, was purchased by the Royal Navy to sail to Tahiti and acquire breadfruit plants. The plants were to be transported to the West Indies where they would be grown as an inexpensive food source.

The tale of *Mutiny On The Bounty*, is the story of the mutiny that occurred on the HMS Bounty on April 28, 1789 against Captain William Bligh. What many do not know is that the Bounty contained a greenhouse. Before her voyage, she was refitted to accommodate the breadfruit she would transport. The renowned British botanist and naturalist Sir Joseph Banks designed many of the changes to accommodate these live plants. In fact, he supervised the modifications to the great cabin, essentially turning it into a floating greenhouse that contained racks for potted plants and a floor that was covered in lead. The scuppers (drains) at the forward corners of the cabin were used to divert excess fresh water from the pots into barrels that were kept below deck to collect the water runoff for reuse.

One thing remains clear, where greenhouse gardening was once the domain of scientists and the wealthy, it is now available to us all.

The passion many people feel for the environment and its bounties are expressed through a simple, yet satisfying activity: gardening. Those who love to grow things yearn for the basic pleasures of planting and caring for flowers, herbs, and vegetables in a garden.

But for many people who live in areas of the country that experience harsh winter climates, gardening year-round is impossible. Because the seasons can limit a gardener's efforts, homeowners can delight even more in a greenhouse, which offers an almost year-round growing environment.

A greenhouse is a structure, usually enclosed by glazed glass or plastic. This covering helps the greenhouse to maintain a comfortable temperature throughout the year. Because it provides favorable climactic conditions for growth, a greenhouse environment supports all kinds of plants, vegetables, flowers, and fruits — literally the bounty of nature enclosed. Best of all, it gives you, the gardener, a good measure of control over the greenhouse environment.

A greenhouse can be built in a variety of shapes and sizes, and often becomes a unique expression of the person who tends to the plants and herbs housed within. Depending on what part of the country the greenhouse is located, there are a variety of plants that can be grown.

Is the gardener growing tomatoes for his or her family? Or, is he or she planting seedlings to be used in a garden next spring? Whatever plants it will shelter and nurture, a greenhouse is a haven for those people who feel compelled to sink their hands into the earth and yield from it all manners of natural wonders.

In this book, you also will become familiar with different types of greenhouses, their basic features, and how to determine which one will best fit your gardening needs. You will discover how to select the best site on your property to build a greenhouse, as well as how to plan your greenhouse, what tools you will need, and how to build it safely. You will become familiar with the step-by-step instructions for building a basic greenhouse from a kit, as well as with the methods for a conventional, do-it-yourself project. You will learn how to build a foundation for a greenhouse, as well as the sidewalls, roof frames, end walls, windows, and doors, as well as how to attach the covering. Finally, you will discover some tips for successfully managing your greenhouse once it is built.

The amount of money and time that is devoted to building a greenhouse can vary greatly from structure to structure. Much depends on the size of the structure and the extent to which it will be customized. Other issues, such as type of greenhouse, location, and climate will also have a significant impact.

It should be noted that the structure of the book is a bit different from others you may have read about building greenhouses. The book begins by explaining the history of greenhouses, followed by the types of greenhouses that you can build. This is normal enough, but then we delve into finding where to put your greenhouse, as well as the foundation to lay down. It may seem odd to talk about building a foundation and then go right into building plans for greenhouses in the following chapters, but there is a method to this madness. The foundation for your greenhouse will be the same for pretty much any greenhouse that you build. A concrete foundation can work for a hoop greenhouse, A-Frame greenhouse, and even a lean-to greenhouse. However, so will a dirt floor, gravel foundation, or wood foundation. So it was important to get the foundation part of the greenhouse out of the way as we did not need to repeat ourselves over and over through the chapters about building a foundation.

The following chapters — after learning the basics of greenhouses — are all about putting together greenhouses, including various plans that you can use. After that, you will learn about greenhouse maintenance, extra accessories and giving your greenhouse its own unique flair.

Photo courtesy of Juliana America, LLC.

Chapter 1

The Roots of Greenhouse Building

> *"A garden is a grand teacher. It teaches patience and careful watchfulness; it teaches industry and thrift; above all it teaches entire trust."*
>
> — Gertrude Jekyll, British garden designer

The art of growing plants, fruits, and vegetables beneath a glass covering is not new. Archaeologists have uncovered evidence that the ancient Romans experimented with growing fruits and vegetables in simple greenhouse-like structures that were covered with thin sheets of mica, a mineral that offered limited amounts of transparency. Gardeners for the Roman emperor, Nero, for example, constructed the first known greenhouse and used mica to fashion its windows during the 1st century AD.

Historic greenhouse in Kassel, Germany.

Historical Roots

The concept of the greenhouse, as was stated above, originated with the Romans, who wanted a way to grow plants in environmentally controlled areas. Tiberius, a Roman Emperor, enjoyed a cucumber-like vegetable on a daily basis and that meant that the Roman gardeners needed to find ways to grow the plant and have it on the table for him no matter the season. To accomplish this, the gardeners would plant the cucumbers in wheeled carts placed in the sun every day, and then brought back in at night where they would stay warm. The cucumbers would then be stored under frames glazed with oiled cloth.

The greenhouse as we know it first appeared in the 13th century in Italy so that the exotic plants that explorers brought back from tropical areas could be grown and harvested for people who wanted these plants in their homes. This 'botanical garden' concept quickly became popular and spread throughout Europe and into England. However, these greenhouses required plenty of work, including having to close them up at night and winterize them, as well as achieving balanced heat within the greenhouse.

The first practical, modern greenhouse constructed was in Leiden, Holland by the French botanist Charles Lucien Bonaparte to house his medicinal tropical plants. From this point on, until the 19th century, greenhouses were found only on the estates of the rich and in universities, thanks to the growth of botany during the Age of Discovery.

By the early 17th century, indoor gardens were being attempted in Northern Europe. French, British, and Dutch explorers who sailed the world found all types of exotic plants, flowers, and fruits. They brought home varieties of plant species, such as citrus trees, that required temperate growing conditions.

In Britain, these explorations coincided with the invention of the **orangery**. An orangery was an unheated structure with a solid roof and glass walls. It was used to house orange and lemon trees and protect them from the harsh winter climate. Typically, an orangery was built against a brick wall. The roof was sloped and the remaining three walls were constructed out of glass.

In the summer, the citrus trees and plantings would be moved outside to gardens, and the orangery could be used for parties and other social functions. It is thought that the word greenhouse itself entered the English vocabulary sometime in the late 18th century. As the use of an orangery became more popular, people began to notice the green color of the plantings and vegetables that were able to grow inside these buildings despite the rigors of a cold, English winter. The term greenhouse was born.

The first large conservatories began to appear around this time, and for the most part, they were used by scientists for research, particularly in the pursuit of medicinal remedies. During the 19th century, as the Industrial Revolution rolled full steam ahead throughout Britain, the aristocracy, who not only kept exotic plants and trees in these structures but entertained in them as well, coveted greenhouses and conservatories. Even in modern times in the United States, it is fairly common to host special events in large conservatories, taking advantage of their resplendent beauty.

In the United States, one of the earliest known greenhouses was built in Boston for a well-to-do merchant named Andrew Faneuil. Faneuil used his greenhouse to grow fruit during the winter. It is also widely believed that the first president of the United States, George Washington, also built a greenhouse on his Mt. Vernon property to grow pineapples.

In 1845, the glass tax was abolished in England, as more and more greenhouses and conservatories were being constructed. An enormous glass building called the Crystal Palace in London was constructed for The Great Exhibition, which was the first in a series of World's Fairs in 1851. It helped popularize the concept of the greenhouse for many individuals as they saw how the inside of the Crystal Palace kept warmth in compared with outside, and how well plants grew within it.

During World War I, heating costs, and a shortage of skilled gardeners in Britain, curbed the popularity in greenhouse construction. It was not long after the war, however, before greenhouses began to take root again, staging a comeback in popularity. Over the subsequent decades, advances in technology, and the development of different metals and materials, such as aluminum and galvanized steel, lightweight polycarbonate, acrylic, and polyethylene glazing, combined with new designs and lower costs to make greenhouses a desirable and much more affordable investment for more people interested in greenhouse gardening. No longer did a gardener have to own an enormous property or be particularly wealthy to enjoy the pleasures of a greenhouse. As with many commercial enterprises, affordability played a large role in popularizing the use of greenhouses. If these structures could be used to grow profits as well as fruits and vegetables, then the sky was the limit.

By the early 20th century, greenhouses were becoming more popular. Farmers intent on improving productivity began to use green houses commercially. In the United States, for example, some farmers utilized greenhouses to grow cucumbers and other produce for sale.

As the commercial greenhouse market continued to blossom thanks to cheap plastic materials in the 1960s, so too did the hobby greenhouse market. Again, developments in technology and more affordable material helped the greenhouse movement grow, and with the push of eco-friendly building materials, plastic, and aluminum, greenhouses are available for a fraction of the cost of earlier versions. They can be purchased in kits. Also, the development of all kinds of accessories, such as fishponds, spas, and hot tubs has flooded the market and is attracting more and more newcomers to the delights of greenhouse gardening.

Savvy homeowners also know about other advantages of having greenhouses on their properties. In a competitive real estate market, a greenhouse can increase the long-term value of a property, because it permits a homeowner to significantly improve the curb appeal of his or her property by improving its appearance. Growing your own plants and trees for landscaping your property can save money, too. Instead of having to rely on a gardening or landscape store for everything, the greenhouse gardener is able to be more self-reliant.

Provided a greenhouse is properly situated, well maintained, and covered with an appropriate glazing, it will provide a long-lasting, healthy environment for all kinds of plants, flowers, fruits, and vegetables. It can be a beneficial environment for a gardener to devote time to his or her craft. After all, the growing environment can be made available nearly all year long, leaving plenty of opportunity to improve his or her gardening skills. There is much more time for practice to actually make perfect, but you must start by understanding the purpose and function of a greenhouse.

Uses for a Greenhouse

One of the great pleasures of having a greenhouse is being able to grow plants, flowers, fruits, and vegetables almost any time of the year. Because a greenhouse gardener is able to take control of the growing environment,

it is, to a degree, a little like bending some of Mother Nature's rules. Depending on which part of the country a greenhouse is located, it is possible to grow vegetables, fruits, and flowers all year long.

Greenhouses provide a great opportunity for families to spend time together. There are a wide variety of gardening activities that all members of the family can participate in.

Greenhouses are used primarily for growing plants, vegetables, herbs, and fruits.

Some gardeners use their greenhouses to extend the growing season. For example, they might seed plants in the early spring and protect them from fall frost. Others might use their greenhouses to grow flowers and plants for a majority of the year. When it comes to plants and flowers, there is a huge variety of plant life that are favored by greenhouse owners. Some gardeners like to grow Chrysanthemums, Impatiens, and Begonias. Others like to raise Sedum and Delphiniums.

Other gardeners prefer to grow vegetables, fruits, and herbs throughout the year. These can be grown in planters, pots, greenhouse beds, grow bags, and hydroponic gardens. A hydroponic garden is a method of cultivating plants using a nutrient solution instead of conventional soil. These also require plenty of sunlight, proper irrigation, fertilizer, and adequate air circulation. Tomatoes, grapes, and strawberries, as well as dwarf varieties of fruit trees are just a small sampling of the fruits that can be grown in a greenhouse. Again, though geographical location and climactic conditions are important factors for operating a greenhouse, so much of greenhouse gardening is based on personal preferences.

Tomatoes being grown in a hydroponic greenhouse.

Besides being able to exercise more control over a growing environment, there are other advantages to having a greenhouse. For example, it can be used to grow not only native and exotic flowers nearly year round, but vegetables and fruits that are free of pesticides and, therefore, healthier to eat. In addition, a greenhouse can provide you with a supply of herbs, grown with your own hands and skill, to spice up a treasured dinner recipe. You also can grow and enjoy varieties of colorful flowers to decorate a table or mantel, or to present as a gift on special occasions.

The pleasures and advantages of owning a greenhouse extend well beyond tending to an indoor garden. For many people, a greenhouse offers a respite from the rigors of an increasingly hectic pace of life. The demands of raising a family, as well as the struggles that come with a busy career, can all melt away while you work in your greenhouse.

Imagine a space that is green and colorful and warm inside, even as the worst winter weather howls just beyond the door. This is your greenhouse. It is another world. Inside, there is peace and quiet. Perhaps soft music plays, a cup of tea awaits, and the pleasant scent of exotic flowers fill the air.

Just the simple act of stepping inside and spending a few minutes to plant seedlings, prune a branch, or check on tomatoes is a restorative tonic to the demands of the outside world. A greenhouse is able to provide the space and time in which gardeners can both lose and find themselves all in the same moment.

Here are a variety of fruits and vegetables that can be grown in a family or neighborhood greenhouse.
Photo courtesy of Juliana America, LLC.

What a Greenhouse Does

Essentially, a greenhouse utilizes just a few basic scientific principles to maintain its warm interior temperature. Sunlight is the primary source of heat for a greenhouse. Sunlight passes through transparent materials, such as glass or clear plastic. When it hits an opaque, or less transparent surface, some of that light is transformed into heat. The darker the surface, the more light is transformed into heat. As the sun's energy heats the air inside the greenhouse, the humidity level is elevated within the structure.

Although heat will remain inside a building that is mostly comprised of glass or plastic, there is always some loss of heat. Therefore, most greenhouses still require additional heat sources. The goal is to trap energy inside the greenhouse to comfortably heat the plants, the ground, and soil inside it. Ideally, the air near the ground is warmed and prevented from rising and dissipating too quickly.

In order for plants to flourish, a greenhouse must provide the appropriate amounts of light, humidity, and warmth. When plants are watered, the humidity inside a greenhouse increases. Humidifiers, evaporative coolers, and misting systems are used to raise the level of humidity. *This will be discussed in Chapter 10.* Because most plants require a stable environment, the key to operating an effective greenhouse is being able to maintain a reasonably stable climate to support them.

Maintaining the right temperature and humidity level, however, is just one part of maintaining successful greenhouse harmony. It is also very important to sustain appropriate lighting conditions, the amount of water plants receive, and healthy soil conditions. If a greenhouse environment goes awry, plants and herbs could dry out and perish. Also, without adequate environmental control and harmony, insects can invade the greenhouse and cause damage to plants and vegetables. Ventilation, heating, and cooling systems, as well as misting systems, must all be taken into careful consideration when planning and building a greenhouse.

CASE STUDY: THE GREENHOUSE NECESSITY

Kate McElhinney
Plant with Purpose
Marketing Coordinator
4903 Morena Blvd. Suite 1215
San Diego, Calif. 92117
www.plantwithpurpose.org

Many individuals have the necessary tools to construct a greenhouse, but there are those who do not. It may seem unfathomable that there are people who do not have the necessary resources to build a green house, but Kate McElhinney sees individuals every day who live with this reality. McElhinney is marketing coordinator for Plant with

Purpose, which is an international, environmental organization that transforms lives in rural areas where poverty is caused by deforestation. For more than 25 years, Plant with Purpose has provided lasting solutions to heal the relationship between people and their environment by planting trees, revitalizing farms, and offering loans to create economic opportunity. "We work in six countries, including Haiti, Mexico, the Dominican Republic, Mexico, Thailand, and Tanzania," she said.

McElhinney said the easiest type of greenhouse the poor can make uses the resources that they have available to them. "In one of the communities we work with Oaxaca, Mexico for example, the people use locally grown wood and local labor to construct their greenhouses," she said. "We provide them with the training to maintain and use the greenhouses." This way the poor are able to maintain their structures. The best greenhouse is one that can be made at a relatively low cost with easily available resources, and one that meets the objectives of the community.

A greenhouse can help to increase crop yields because the temperature can be regulated, pests can be controlled, and the people are also able to grow foods that might not do as well on the outside. For example, in Nuxino, Mexico farmers have been successful in growing tomatoes because the greenhouse creates a warmer environment for them to thrive in. "In our experience, greenhouses help the environment because they use less water, or the water is more regulated with a drip system. The farmers use locally grown wood to construct the greenhouses, as oppose to cutting down forests. Also, these communities typically only own small portions of land, so greenhouses allow them to produce higher yields, which mean that there is more land left to use for other means, such as farming or tree planting.

"The community of Loma Chimedia in Nuxino, Mexico is one example of a group we are working with who have seen tremendous improvements in the quality of their lives through the use of greenhouses," said McElhinney. This group started growing tomatoes, which have flourished in the greenhouse because of the warm temperature. The farmers have done well selling the tomatoes, and have been able to thus feed their families and send their children to school. According to McElhinney, their quality of life has dramatically improved.

To learn more about Plant with Purpose and how you can help, please visit **www.plantwithpurpose.org**.

Royal Greenhouses at the Palace in Laken, Brussels, Belgium.

Greenhouse Basics: Conduction, Radiation, and Convection

Heat always flows from a warm location to a cold one. It is transferred in one of three ways: conduction, radiation, and convection. Conduction means heat energy passes through a material, such as a nail. Conduction refers to the transfer of heat, or thermal energy between molecules. Conduction occurs in all forms, such as solids and liquids. Typically, metals make better conductors than wood.

Radiation, on the other hand, refers to the transfer of heat even if there is no solid medium. Radiation is any process in which energy travels through a medium or space and is absorbed by another body. Have you ever placed your hand near a candle and felt the heat that radiated from the flame? That is a radiating heat. An object that is hot gives off light that is known as thermal radiation. The hotter the object, the more light it emits.

Convection refers to the movement of molecules within liquids or gases. Convection is the heat transfer within an environment. If warm air flows out of a room, then cold air will enter the room and replace it. This is a prime concern when it comes to maintaining a healthy environment inside a greenhouse.

Learning these processes will help you understand another process called the greenhouse effect, which occurs when solar radiation passes through a transparent object and heat is absorbed inside an enclosure. It is also important to understand the three fundamental ways heat is transferred, because then you can determine what kind of heating system you will need inside your greenhouse in order to keep your plants healthy during temperature drops and increases.

Onward to Greenhouse Types

In the past few pages, you have learned about the history and the science of the greenhouse. Though knowing where greenhouses have come from and the evolution of them is significant, the science is an important concept as well. However, if you do not quite understand how a greenhouse works, as in how convection helps keep your plants growing in the greenhouse, do not feel that you are behind in this book. Understanding the different types of greenhouses that you can build is most important. This is crucial because your skill with building things, the space you have in your yard, and the amount of plants you want to grow will all dictate just what type of greenhouse you want to build.

Types of Greenhouses

Greenhouses can be divided into two basic types: attached and freestanding. As its name suggests, an attached greenhouse is a lean-to structure that is attached to a house or another building, such as a garage or main house. A freestanding greenhouse stands by itself; virtually anywhere on a property its owner decides it should be. A freestanding greenhouse can be as small or as large as the property's space will allow.

Attached Greenhouses

It is helpful to think of an attached greenhouse as a room added on to a house. It can be as small as a bay window or literally the size of a room with a foundation, wooden framing, and glazed windows. An attached greenhouse presents several advantages, as well as challenges, for a gardener. One big plus is that, because it usually shares a wall with a main dwelling, this type of greenhouse is able to receive heat from the main house. Considering how expensive heating costs can be, especially in a geographical location that experiences cold winters, this type of greenhouse can save money. The shared wall also means there is one fewer wall that has to be built. Another advantage an attached greenhouse offers is its proximity to the main dwelling's utilities. What could be more convenient than having a greenhouse space located close to pre-existing plumbing and electrical systems? It makes tying in to electrical and water sources much easier.

Another advantage of building an attached greenhouse is its ability to provide an added layer of insulation to a house. An attached greenhouse acts as a buffer zone between the walls of the house and the outdoors. And during the cooler months, an attached greenhouse, with its glazing, can help heat a house on bright days, which can also lower the overall heating costs for the main dwelling. Another issue to think about is access to the greenhouse. During the winter, when snow is piled high outside, entering an attached greenhouse is going to be much more convenient than shoveling your way through piles of snow to reach a freestanding greenhouse.

On the other hand, an attached greenhouse can be subject to, or limited by, the home's design. Most homeowners want an attached greenhouse to visually blend in with their home. There is not much curb appeal when a greenhouse sticks out like a sore thumb. It helps if the attached greenhouse space can incorporate those materials that will best blend in with the main house to maintain a consistent, pleasing aesthetic.

Depending on the design of the main house, a low-sloped greenhouse roof or small exterior wall can make that architectural harmony difficult to achieve. Also, because of the limited available space the greenhouse may provide, it could be difficult to increase the size of the plants you plan to grow. There is also another issue to consider when planning to build an attached greenhouse. It will typically be humid inside and humidity can seep into the walls of a house and cause rot.

This is an example of an attached greenhouse.

Lean-to Greenhouse

One of the most common types of attached greenhouses is the lean-to. Typically, this three-sided greenhouse is attached to a home's southern wall for maximum exposure to sunlight.

One of the advantages of a lean-to is that its roofline can be adjusted to match the roofline of the main house. As mentioned earlier, because the lean-to is an attached greenhouse, it has the advantage of sharing the main dwelling's wall, which also means it receives heat from the larger building. A lean-to makes it more convenient to tap into the home's plumbing and

electrical systems, too. This is significantly less expensive than heating a freestanding greenhouse and providing it with electrical power. However, because a lean-to is typically smaller than a freestanding greenhouse, it can overheat more easily without adequate ventilation.

As you learned earlier, another potential downside is the need to ensure that the attached lean-to greenhouse is constructed with

This is an example of a lean-to greenhouse.
Photos courtesy of Juliana America, LLC.

materials that visually complement the main dwelling. Remember curb appeal. An attached greenhouse with a high, steep-sloped roof might look awkward when directly attached to a house with a low-sloped roof.

Another possible disadvantage is that the lean-to's limited space could make it challenging to grow larger plants, such as fruit trees. There also is a potential safety concern here: A lean-to should never be attached to a building where snow could slide from a higher level, such as a second or third floor roof, onto the greenhouse. A snow-slide could damage the greenhouse covering, or even injure someone

This is an example of a lean-to greenhouse.
Photos courtesy of Juliana America, LLC.

inside the building. This means it could be a potential problem to place an attached greenhouse directly beneath the steep-sloped roof of a house if heavy snowfalls are common to the geographical location.

Window-mounted attached greenhouse

This is one of the most basic and most common types of greenhouse spaces. A window-mounted greenhouse is typically small in size. Because it is attached to the window and it offers limited space, a window-mounted greenhouse tends to heat up quickly. During the summer, it will probably require blinds, shutters, or curtains to help protect the plants from overexposure to the sun, keeping them from overheating. One way to help prevent plants from overheating is to place a thermometer in the greenhouse space and monitor it. It is advisable to build a window-mounted unit so that the original window in the house will not close. That way, the window-mounted greenhouse will not overheat when it is warm outside, or, conversely, cool down when the outside temperature drops. Blinds or shutters can also be closed if an abundance of sunlight causes the space to overheat.

A window-mounted greenhouse has advantages because it provides the means to practice your gardening skills on a more modest, less expensive scale than larger greenhouses. It also allows you to build on those skills, incrementally, without having to worry about expanding the space. Another advantage of a window-mounted greenhouse is its ability to provide additional heat and humidity to the room it faces. This type of greenhouse space is suitable for starting seeds, as well as for growing a variety of flowers and herbs. The greenhouse effect tends to multiply in a small, sunlit space. Remember, a greenhouse collects sunlight. That sunlight heats the air and elevates the humidity around the plants. Ventilation releases heat and exchanges carbon dioxide for oxygen.

Even-span

An even-span, or traditional-span, is another common type of greenhouse. It can be attached to the house or it can be a freestanding structure. The

main characteristic of this type of green-house design is the traditional even-span gables. A **gable** is the triangular part of the wall between the edges of a sloped roof. This type of green-house typically has vertical sidewalls and

This is an example of an even-span, freestanding greenhouse.
Photos courtesy of Juliana America, LLC.

an even-span roof with plenty of headroom along the center of the structure. When it is attached to a building, like a house or garage, the larger building shelters an even-span greenhouse.

The sidewalls are usually about 5 feet (1.5 m) in height. The central roof ridge is generally 7 to 8 feet (2.1 to 2.4 meters) high. Also, an even-span green-house will often have floor-to-ceiling glass windows. The frames for this type of greenhouse are often made from alu-minum. Roof vents are hinged on one side and can be opened with arms or levers. Many gardeners use automatic vents in their greenhouses. This type of powered vent is controlled by a ther-mostat that opens the vent depending on the need for ventilation.

This is an example of an even-span, freestanding greenhouse.
Photos courtesy of Juliana America, LLC.

Freestanding Greenhouses

You learned that an attached greenhouse is typically attached to a house or larger building. A freestanding greenhouse, as its name implies, is situated apart from the main dwelling, or any other buildings. It is common for freestanding greenhouses to have their own sources of water and electricity, although there are some that do not. Because all four of its sides are open to sunlight, a freestanding greenhouse can be located almost anywhere on a property, provided it is built on fairly level ground and is positioned for maximum exposure to the sun through the different seasons. Typically, a freestanding structure costs more to build. One of the simplest and least expensive is a basic hobby greenhouse, with a polycarbonate covering. This type of freestanding greenhouse can cost approximately $600. Larger, more sophisticated freestanding greenhouses can cost up to $10,000 or more.

One reason for the significant price difference is the size and design of the greenhouse. Depending on its style, the greenhouse might require a concrete foundation, which means more supplies, construction work, and time. However, there are freestanding greenhouse kits that provide a much less expensive structure and do not require a concrete foundation and customized components. Typically, a greenhouse kit contains all the components needed to build the structure with maximum ease. For example, a hoop style greenhouse kit contains the following materials.

- Galvanized steel hoops
- Steel hoop support stakes
- Steel hoop clamps
- Steel door hinges and handles
- Hook and eye door locks
- Turnbuckles
- Poly-tubing joiners
- 6-mil UV resistant greenhouse film
- Batten tape

- Wire
- Velcro Tape
- Fasteners
- Building plans

This is an example of an even-span, freestanding greenhouse. Courtesy of Juliana America, LLC.

Unlike an attached greenhouse, however, a freestanding structure is more open to the elements, so it typically requires sturdier framing and glazing to stand up to harsh weather. Also, because all four sides are exposed to the weather, heat loss is a critical consideration, especially during the winter months. Because electrical power and plumbing must be provided, this type of greenhouse can be more expensive than an attached greenhouse.

Cold frames and hotbeds

Freestanding greenhouses come in a variety of shapes and sizes. Two of the most simple and least expensive types are cold frame and hotbed green-

This is an example of an even-span, freestanding greenhouse. Courtesy of Juliana America, LLC.

houses. It will help to think of them as mini-greenhouses. A cold frame greenhouse is basically just a ventilated box placed on the ground that is covered with glass or plastic. One of the chief advantages of a cold frame greenhouse is that it is relatively easy to build. The back frame of the box stands approximately twice as high as the front of the box, so the glazing that covers the box rests on a slant from top edge to bottom edge. The box should be tall enough to fit the plants that you want to grow inside it. The frame can be constructed of plastic, wood, or brick.

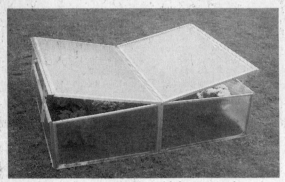

This is an example of a cold-frame greenhouse made out of polycarbonate.
Photos courtesy of Juliana America, LLC.

One main consideration is that the box must prevent drafts from entering. Also, the soil must be fertile and free of weeds. A cold-frame greenhouse should be positioned so that it will receive maximum exposure to sunlight during winter and spring. The ideal location for this type of greenhouse is against the wall of another building, if possible, so that it is protected from the wind. Ventilation is critical to ensure that the environment remains cool. A cold-frame greenhouse is ideal for starting a variety of plants and some flowers, such as geraniums.

A hotbed greenhouse is similar in design and construction to a cold frame, except that it utilizes manure, which releases heat as

This is an example of a cold frame greenhouse made out of polycarbonate.
Photos courtesy of Juliana America, LLC.

it decomposes. The manure is deposited and set in the bed frame below ground. Then a layer of soil is placed over the manure. A hotbed greenhouse also must be ventilated to prevent overheating.

This is an example of a hot-bed greenhouse built out of bricks.

Quonset

A Quonset greenhouse, or hoop-house, as it is often called, is a popular style of freestanding greenhouse because of its design, which covers the largest area of ground for the least amount of money. Check out your local garden center. If there is a greenhouse on the property, chances are it is a Quonset.

This is an example of a Quonset greenhouse.

Typically, a Quonset greenhouse is made with galvanized steel arches covered with plastic. A Quonset can be just about any length and size. The length of the structure depends on how much available space there is on which to build. The basic design for a Quonset construction calls for a hoop, often made from PVC pipe, to be installed every 4 feet along the length of the structure and a spacer, or space bar, placed every 10 feet. The hoop is what gives the Quonset a distinctive shape.

Many Quonset greenhouses are covered with construction-grade polyethylene. This material, however, is not UV stable, which means that over time, it can degrade in sunlight. Also, because of its light weight, this type of covering can be susceptible to damage by high winds. Some Quonset-style greenhouses are covered with polycarbonate that is bent and shaped

This is an example of a Quonset greenhouse.

to fit the curve of the roof frames. Although it is a more expensive option, this type of glazing lasts much longer than less costly plastic coverings. A Quonset greenhouse will allow a gardener to extend a growing season for two or three months.

Rigid Frame

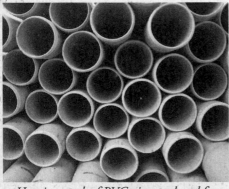

Here is a stack of PVC pipes gathered for constructing a new greenhouse.

Greenhouse frames for freestanding structures, no matter what their shapes or sizes, can be constructed with a variety of materials. Aluminum and wood are probably the most popular materials used for framing; however, there are other materials, such as PVC, that greenhouse builders use for their construction projects. PVC is a tough, synthetic resin that is made with polymerizing vinyl chloride.

Wood framing is attractive, and it is also sturdy, and depending on the wood type, rot-resistant. Many greenhouse gardeners prefer to use western red cedar or redwood because of this material's inherent ability to resist damage caused by rot.

Another option greenhouse owners favor is pressure-treated wood. This type of material has had a chemical preservative forced into it, which makes it tough and resistant to insects and water damage.

If aluminum is the preferred choice of material, there are several significant advantages. First, aluminum requires very little maintenance. Also, not only is it very strong, it is lightweight and

Here is an example of condensation that has built up on aluminum.

enjoys a longer lifespan than wood. Aluminum, which can be color-coated, is sold with the majority of rigid frame greenhouse kits. Typically, these kits are not difficult to assemble, and parts often include some pre-drilled holes. One of the disadvantages of aluminum, however, is that it will lose heat more rapidly than wood, which can make a greenhouse with aluminum frames more expensive to heat. Also, condensation can be a problem when aluminum frames are used. Condensation occurs when water vapor contained in the air is converted by changing temperatures and becomes droplets of water that gather on the aluminum.

Still, another material used in rigid frame greenhouse construction is galvanized steel. This type of framing is typi-

Here is an example of galvanized steel that has gone through a chemical process to keep it from corroding.

cally found in larger commercial greenhouses because galvanized steel is very strong and durable. However, steel is much heavier than wood or aluminum, and thus more difficult to handle.

Post-and-Rafter

This is an example of a barn being built that utilizes the post-and-rafter or pole barn style construction. Please note there are no sills or a foundation.

A freestanding greenhouse also can be framed with post-and-rafter construction. This is an easier method than a standard platform construction because, instead of being built up from the sills, posts are planted into the ground and the rest of the architecture is hung on them. Traditionally, post-and-rafter construction utilizes fewer components, but require larger members. After the posts have been installed into the ground, the tops are cut off so that all the posts are level.

A-frame

A well-constructed A-frame greenhouse is a study in simplicity. Compared to other types of greenhouses, an A-frame is relatively easy to build. The A-frame takes its name from its simple, unique design. It is comprised of two flat sides that lean against each other, and resembles the shape of a capitalized letter "A." This type of greenhouse has two end walls in between the flat slides.

To build an A-frame greenhouse, the walls are first constructed in sections on the ground. Then the walls are raised, positioned, and finally joined.

The roof slope can vary from greenhouse to greenhouse, but generally it is steep enough to efficiently shed rain and snow. Typically, the A-frame is secured to the foundation. Corrugated polycarbonate or acrylic is often used as the glazing for the greenhouse. These glazing materials can be attached to the A-frame's structural beams with relative ease. Another advantage an A-frame greenhouse offers its owner is the convenience with which doors and shelves can be installed at the ends of the building. Because the A-frame is designed with greater height along the center of the building, it is more convenient to grow taller plants and trees in this type of greenhouse.

The downside, however, are its sloped walls, which generally allow less headroom than other types of greenhouses. To get more headroom out of an A-frame, the frame itself can be installed on short walls. The additional height this provides will allow for a wider roof angle and more headroom. Another disadvantage to the A-frame is that the amount of floor space is small in relation to the height of the greenhouse. This can be a problem in winter because heat rises, and in an A-frame, it will rise far above the floor and the plants that need the consistent warmth. A fan installed in the ceiling reduces the heat loss by pushing warm air back down towards the floor. Another issue to think about is that an A-frame also has wasted space at the foundation. There is little room there, so it is not ideal for low-growing plants.

Conventional, vertical-sided

Here is an example of an even-span vertical-sided greenhouse.
Photo courtesy of Juliana America, LLC.

This type of greenhouse is one of the most widely used and one of the easiest to construct and maintain. The sides are vertical, which generally makes installing shelves and benches inside relatively uncomplicated. The roof is sloped toward a central ridge, which allows plenty of headroom along the central aisle. The door is typically located at one of the two peaked ends. A sidewall in a conventional greenhouse is often about 5 ft. (1.524) in height. The central ridge is typically 6 feet (1.82 meters) to 8 feet (2.43 meters) in height. A ridge is the horizontal line on a roof that is formed by two sloping planes.

Because this type of greenhouse has glazing on all four sides, there is one major drawback. The glazing on the north side of the building will allow heat to escape and may not receive enough sunlight to make up for that heat loss. A way to remedy that loss of heat, however, is to install insulated panels on the north side of the greenhouse during the winter months to help minimize the loss of warmth.

Greenhouse kits

Thanks to mass-production of supplies and lightweight/easy-to-use materials, almost any kind of greenhouse comes in a kit. Check greenhouse supply catalogues for choices in size, materials, and coverings. Most hobby greenhouses have rigid or flexible plastic coverings. Kits with less prefabrication are typically less expensive but require more labor to assemble. A kit should include hardware, instructions, and a telephone number for customer support. When you open a kit, a good rule of thumb is to lay out all the components and review the packing list to ensure you have all the parts.

Summary

Greenhouses have come a long way since being nothing more than cucumber carts during the age of the Roman Empire, to being massive buildings used only by the wealthy. Now, you can find a greenhouse anywhere and they range from the very large to the very small, to meet all the needs of avid gardeners.

As you can see, there is a greenhouse to accommodate almost any level of gardening skill and need. Although the type and style of greenhouse is up to the gardener who plans to work inside it, there are important issues to consider before construction takes place. Geographic location, climate, terrain, types of plants that will grow, and available utilities are just a few of these considerations that you will need to address in order to choose the type of greenhouse that will best help you to enjoy — and practice — your love of gardening on a regular basis.

Elaborate greenhouse that is in the Kuskovo estate in Moscow.

Chapter 2

Choosing the Right Greenhouse for You

> *"For myself I hold no preferences among flowers, so long as they are wild, free, spontaneous. Bricks to all greenhouses! Black thumb and cutworm to the potted plant!"*
>
> — Edward Abbey, American author and essayist

With all of the different types of greenhouses discussed in Chapter 1, the decision to build your own greenhouse, rather than rent a space that is already available, may seem like a complex one. But, the type of greenhouse you choose will depend on several factors — discussed in this chapter — so your choices can easily be narrowed down. For instance, have you decided what kinds of plants you would like to grow? Think about the topography of your property like. Is it level or hilly? Also, what is the weather like where you live? Is it temperate all year long or are the winters extremely cold and snowy? All of these things will limit, or permit, the type of greenhouse you will be able to build.

Climactic Conditions

There are a number of important factors to take into consideration when deciding which greenhouse will best suit your gardening needs. Most importantly, what types of plants or flowers do you plan to grow? Do you want to cultivate tropical flowers or tomatoes? Orange trees or lettuce? Varieties of plants and flowers have different climactic needs. Another consideration to think about is the climate where you live. Do you live in a region that gets heavy snowfall? Or, is your home located in a more temperate area? Seasonal weather conditions will have an impact both on the type of greenhouse you build and the plants you want to grow inside it. Another consideration is space — how much land will be set aside for the greenhouse?

Seasonal weather conditions are always a concern for greenhouse gardeners. Not only can the climate dictate what type of greenhouse to build, it can play a critical role in determining how a greenhouse will function after it is built from one season to the next. For example, in a cold-weather climate that frequently experiences heavy snowfall, a steep-slope roof might be ideal because it can prevent the build-up of ice and snow by efficiently shedding them. In a location that sees less snowfall but gets cold, strong winds, a sun pit, which is a greenhouse with the majority of the structure housed below ground, might be the best choice, because it is naturally insulated and requires less heat to operate. The same type of roof might not be practical for a more temperate area.

If you live in a part of the country that frequently experiences high winds, is a freestanding greenhouse your ideal choice? In such a location, a free-standing greenhouse might demand extra bracing. Instead of building a freestanding structure, it may be more practical to construct an attached greenhouse that is partly sheltered from the elements by the main dwelling. In rainy parts of the country that see less sunlight, such as the Pacific Northwest, an attached greenhouse with interior access offers protection from the elements, because you can enter it from the inside of your home. Also, utilities, like electricity and water, are literally right next door. In an area where heavy rainfall and snow is common, a steep-slope A-frame greenhouse could be a wise choice, because it sheds them more efficiently than a lower-sloped roof.

It is very important to determine how a greenhouse will be used before building it. Knowing what will be grown in the greenhouse and how the space will be utilized will have a direct bearing on the best type of greenhouse for your needs. What kinds of things will be grown in it? Do you plan to grow herbs or vegetables? Or, will it house plants or fruits? Is the greenhouse to be used all year long or simply to extend the growing season? Unless you live in a warm climate, an unheated greenhouse is really only appropriate during the spring, summer, and fall seasons when the sun will be able to provide enough heat for the greenhouse. During these seasons, the residual heat form the sun in the greenhouse will keep everything nice and warm for the plants. A greenhouse that will accommodate tropical flowers in the midst of a severely cold winter climate will require constant heat. Most vegetables, for example, can grow in a slightly cooler climate, without the humidity many flowers require. A heated greenhouse is more versatile and can accommodate a wide variety of plants. With the addition of heating and ventilation, adequate shade, and water, a heated greenhouse can protect plants from colder weather with a nurturing environment in which to thrive.

It is also important to think about the equipment and materials that will be required to maintain the greenhouse before the construction begins. Again, you will decide what types of plant life you will grow in your greenhouse. If your goal is to simply extend the growing season for lettuce or geraniums, your greenhouse could be a smaller, unheated structure, such as a cold frame. If you are more ambitious, then a larger, permanent building with electricity and plumbing might be more appropriate for your gardening needs.

Temperature Controlled Greenhouses

There are three basic types of indoor growing environments into which plants are grouped: Tropical (also known as hot), warm, and cool greenhouses. These environments were developed to replicate the actual environmental conditions in which these plants and flowers grow. Understanding these types of environments, and the kinds of plant life each supports, will also help you determine which type of greenhouse will best suit your needs. For example, if you want to grow a tropical plant, such as an African violet, your greenhouse should foster a brightly lit, humid environment. On the other hand, a greenhouse with a cool environment is more suitable for growing vegetables.

Hot greenhouse

To be truly effective, a tropical greenhouse must maintain a higher temperature during the winter: a minimum of 65 degrees (18.3 C) at night and 75 degrees (23.8 C) to 80 degrees (26.6 C) during the day is recommended. This environment is appropriate for growing more exotic tropical plants, such as orchids and bromeliads. To maintain those temperatures in a tropical greenhouse during the winter, however, you will have more expensive heating bills.

Warm greenhouse

The average temperature for this type of greenhouse usually is 55 degrees (12.7 C). During the cold winter months, a warm greenhouse should be heated at night. In the winter, with its shorter days, daylight can be supplemented with artificial lights. A warm greenhouse environment is good for growing flowers, fruits, houseplants, and vegetables.

Cool greenhouse

A cool or unheated greenhouse is ideal for leafy green vegetables, or for starting seeds. They can grow at temperatures ranging from between 40 degrees (4.4 C) and 45 degrees (- 2.7 C). Depending on the severity of the winter climate, a freestanding greenhouse that does not have a heating system can be used to grow vegetables for most of the year. An unheated vegetable greenhouse can be a smaller structure, not much taller than the height of the plants. One of the advantages of an unheated greenhouse is that it does not need much winter light, and thus it does not require thick glazing. During the winter, turn off and drain the water supply in a cool greenhouse.

Specialty Greenhouses

Some greenhouses have not only certain temperature guidelines in order to grow the healthiest plants, but some even have specific altitude gauges required to grow specialty species such as orchids and Alpine plants. These two types of greenhouses enable a gardener to produce the perfect specimen in climates and altitude levels that would not otherwise be able to thrive properly without these conditions.

Orchid greenhouse

Orchid greenhouse

For those gardeners who want to grow orchids, some species need specific growing conditions. For example, though some orchids thrive in a dry climate, others will do much better in a more humid environment. Some orchids need specific temperatures during daytime or nighttime hours in order to thrive. Most will not grow very well if they are exposed to exhaust fumes from heaters or sprays from pesticides. Heating costs can be another consideration. One way to minimize the heating costs for growing orchids during the winter is to build a lean-to greenhouse attached to a house. As you learned in Chapter 1, because an attached greenhouse shares a wall with the main dwelling and can open into it, the heat from the house will help to keep the greenhouse itself warm.

Alpine greenhouse

An alpine greenhouse is best suited for plants normally found at high altitudes where conditions are bright and cool. Plants such as sedum, gentian, and edelweiss do well in this type of environment. This type of greenhouse must be able to provide the plants with a cool atmosphere, yet prevent water or snow from collecting on them. This type of greenhouse typically has a glass roof and open windows with screens to prevent insects from entering the space during the warmer weather. The windows can be louvered to allow air circulation, and, at the same time, prevent rain or snow from gathering on the plants. A **louver** is a window or blind with adjustable, horizontal slats that are set at an angle to let in light and air, but to repel rain, snow, and direct sunlight. If you cannot louver your greenhouse, you can try using another method — fans will also help keep moisture from settling on the plants.

Property Considerations

Now that you are familiar with the different types of greenhouses, it is time to think about where on your property your greenhouse will be located. There is more to situating your greenhouse than simply having available space. Look around your property:

- Is the land fairly level?

- Is there an area that is sheltered from the wind?

- If you plan to heat the greenhouse, will you be able to provide the necessary utilities such as electricity from conveniently located service outlets?

Keep in mind that you will want to orient your greenhouse so that it receives maximum exposure to the sun. That means at least six hours of sunlight every day, throughout the year.

What if You Do Not Have Space?

One problem with greenhouses is that they take up plenty of space. Although you can have a small greenhouse box that can literally sit on an apartment balcony, what do you do when you want to have a larger greenhouse? If you want to grow more food, you are going to need more space. Thankfully, you can look to commercial greenhouses, which you rent from the commercial growers, as a place to grow your food. This is a good solution if you do not have space where you live. In order to find a commercial greenhouse to rent, do the following:

- The first thing you should do is look for non-profit and community gardens. One great thing about community gardens is that they have knowledgeable people who work the gardens and can help you get the most out of your growing. If you talk to environmental non-profits, you will be able to get a list of greenhouse owners who you can contact to rent a space.

Community greenhouses such as these allow families with limited or no space available to have their own garden. Community greenhouses may also give you an opportunity to sell any excess harvests you may have.

- You can place an ad in a newspaper asking to rent space in a greenhouse. You can also put an ad on Craigslist.com and Kijiji. com. State how much space you need, where you are located, and how to contact you.

- Call plant nurseries in your area. Ask them if you can rent space and if they do not rent space, ask them about places that do rent greenhouse spaces.

- Talk with farmers in your community. Many farmers will be happy for the extra cash, and you will not be taking up much space on their land. It is a great idea and it can give you plenty of space to work with.

Where Greenhouses Are Used the Most

Naturally, greenhouses are used heavily in northern latitudes and less and less in southern latitudes. For example, in Canada a greenhouse will extend the growing season by several months. Most areas of Canada can only grow gardens from May to August. This is not long enough for many vegetables and plant varieties. A greenhouse can extend the growing season by several months. This allows for growing from as early as March to October. In southern latitudes, greenhouses are mostly used to provide protection from pests and to give plants protection from pests and adequate humidity in a sheltered setting.

Location, Location, Location

Most people are familiar with the real estate agent's popular credo, "Location! Location! Location!" Well, the value placed in choosing a location to build a home also applies to selecting the site for erecting a greenhouse.

Site selection

The site where you decide to construct your greenhouse will play a big role in its success or failure. Before construction begins, consider the following criteria: your greenhouse will need to be exposed to at least six hours of sunlight each and every day regardless of the type of structure you build.

A greenhouse should always be oriented so that it is exposed to the maximum amount of sunlight possible. Because of the sun's movement across the sky, a greenhouse that is positioned so that the long side of the building faces east to west will capture much more sunlight than a structure that is oriented north to south.

Also, it is important to remember that the sun's position in the sky changes at different times of the year. A greenhouse that is exposed to full sunlight during the summer can be cast in shadows as a result of the sun's lower position in the sky during the winter months. Another reason to avoid building too close to trees is that you want to ensure branches and other debris will not fall on the greenhouse and cause damage, or worse, crash through the covering and injure someone inside the building.

As much as possible, the greenhouse also should be protected from exposure to high winds. In windy areas of the country, a fence or hedge can provide a windbreak for the greenhouse.

The surface on which a greenhouse is built is just as important as where it is constructed. Build on firm, level ground that has a minimal slope to allow for adequate drainage. If water does not allow for adequate drainage, the soil can become hazardous for both people and plants. Left to stand, the damp soil can foster mosquitoes and become a breeding ground for parasites that could harm your plants. If the soil is sandy or very moist, the greenhouse will require a solid foundation.

When choosing a site for a greenhouse, it is important to think about convenience. The closer a freestanding greenhouse is located to the main house, the easier it will be to carry supplies like bags of soil, tools, and supplies back and forth from one to the other. And on those chilly, snowy days during the winter, it also will be less unpleasant to step out of the house and walk a short distance to the greenhouse.

It will also be helpful to locate the greenhouse near the garden. The farther away it is from the garden, the more work you will have to do when it comes time to shuttle supplies, plants, and tools between them. If utilities such as electricity and plumbing are located nearby, it will be much more convenient to connect them to the greenhouse. This also is a cost consideration. The farther away the utilities are from the greenhouse, the more time and money you will need to install them.

Typically, an attached greenhouse is located on the south side of a house. When planning this type of greenhouse, consider the orientation of the building to which it will be attached. A greenhouse with a glazed roof located beneath a house with a steep-slope roof is an accident waiting to happen. Imagine the damage heavy snow could cause if it slid off a steep-slope roof onto a glazed greenhouse covering below.

Geographic Considerations

You have become familiar with the importance of building on level ground with adequate shelter from the wind. It also will be helpful if you figure out how many hours of sunlight the greenhouse can expect in each of the seasons it will be in use. The farther north the location, the less sunlight there will be during winter. Noon is the best time to receive sunlight into a greenhouse. The glazing should be positioned to accommodate the sun's location in the winter sky. This will allow for less reflection and more direct sunlight.

Hilly Area

If you live in a hilly area, it can be hard to build a greenhouse, because you need level ground. You also need to ensure that you put the greenhouse in a spot that will give you the most sunlight, without being shaded by the hills around you. If you live near where there are many hills, you will miss direct sunlight in the morning hours and evening hours, which could cut down your direct sunlight hours to only five to six. In the winter, this is made worse by the angle of the sun being in line with the hills, cutting direct sunlight hours down to one or two.

High Rainfall Area

If you live in a high rainfall area, you will have plenty of water but not much sun. As a result you need to maximize as much sunlight as you can get by having a large surface area for your greenhouse that runs from east to west in order to bring in as much sunlight as possible.

High Temperature Area

If you live in southern latitudes, you are going to get plenty of heat. This means you need to have vents in your greenhouse so that the temperature in the greenhouse does not get too hot. The humidity and heat in a greenhouse, if too high, could actually kill your plants if you are not careful.

Northern Latitude

If you are in the north, specifically Canada and the Northern States, you will not get as much sunlight as other areas of the world. Make sure you have a greenhouse that maximizes the surface area to collect sunlight, while not being too large — otherwise it will be too hard to heat.

CASE STUDY:
THE GREENHOUSE DECISION

Vince Panero
Marketing Director
www.thegreenv.com

Vince Panero is not your typical greenhouse gardener; in fact — for a while — Panero did not consider himself a gardener at all. He simply enjoyed the fruits of his father's labor — a labor that had been in the family for decades. "My father, who passed away in the summer of 2004, came from a lineage of Italian/French farmers," said Panero. He raised a garden all year long, along with rabbits that he raised for meat and manure for his garden, in the Mediterranean climate in the central coast of Calif. Panero said he loved that garden, and would go into the middle of it every night to watch the stars (when the sky was not overcast), and he would attain an amazing level of calm. "That was my heritage, but I travelled and lived overseas so much that rarely did I farm ... not even a simple garden," he said.

Eventually, Panero moved away from the sunny weather of Calif. to the Northwest, where it was much colder. "Then, I lived at an eco-community in Dexter, Oregon with my wife and son called Lost Valley Educational Center, in the middle of 84 acres of pine, Alder, cedar, oak savannahs, and newly planted cascara trees," he said. "Outside our little apartment was a 1,000 square foot space of brambles. I cleared that area, and began planting a garden. I did great, but I always envied the amount of food LVEC gardeners harvested from their 8 foot tall 'hoop green houses'."

Panero said you could walk in; it was very humid due to the constant watering and you could tell that the plants loved it. Panero described it as the Amazon Rain Forrest inside while it was a dry, drought-like summer on the outside. "Part of me wanted that. I moved, and I created another garden," said Panero. "Each time, I was challenged by frosts and my seedlings died due to bad timing — planted too early, or too late."

Like many gardeners and greenhouse enthusiast, Panero could attest to being distracted by life in general, but said gardening was about responding at just the right time, and doing just the right thing to nurture, cultivate, and ultimately harvest. "I did not want to build something I would have to move, so I held out on the greenhouse idea," he said. Some time passed before Panero finally bought a house; thus, giving him the opportunity to follow in this father's footsteps. "I felt I had finally reached a state of permanence, and a greenhouse that would give me more time to start my seedlings, to plant, to harvest, free from frost — a magic thermal fly wheel that would start my season earlier for me and add on weeks so I could keep growing through to early winter," said Panero.

Once Panero had the space required to construct a greenhouse he began researching the specific types of greenhouse that were available. "I started out by looking online for a greenhouse cloth," he said. "Basic old clear plastic degrades in a season, and creates flakes that fall into your soil — and they will end up in the ocean as pollution sooner rather than later." Instead, he said he encountered a 5 to 10 year guaranteed greenhouse plastic when he was living at Lost Valley Educational Center. He managed to pick some up in great condition and still had half a roll to spare. "I used it to create a 6 x 14 greenhouse," he said. "It was hoop style — PVC plastic with rebar in it, and green house cloth clamped over the top. Not 'pretty,' but very affective — and warm and humid."

In order to build his greenhouse, Panero measured off the available area, then located equidistant holes in which the half hoops would rest.

1. He hammered in holes (with a hammer, sledge hammer, and wood spike) 1-inch wide and 1-foot long (it can be a two-by-two or it can be a round branch — it does not matter), then he dropped in 3/4 PVC tubes 8 inches long.

2. He then placed rebar into 1/2-inch PVC pipe, cut them to match in length, bent them over, and created the hoop. Six to eight of these acted as a flexible skeleton.

3. Next, he added a 1/2-inch rebar/PVC spine down the top and diagonally down the side — larger sized zip ties held all in place.

4. Lastly, he created a doorway on the front by pounding two-by-four into the ground and attaching a door frame (made out of two-by-twos). I attached the door frame to the two-by-fours and to two places on the first hoop.

5. Finally, he cut the cloth to make a half circle, clamped it on the end with the door, cut a hole that would swing for the door, then cut a larger piece of green house cloth to feed over the top and hang over the other edge. Clamps (spring loaded or 'green house' clamps) held the green house in place during windy weather, etc. Down and dirty — easy.

Panero said that having a greenhouse has improved his gardening in many ways. "All my asparagus came up earlier and stronger," he said. "My tomatoes also rocketed last year." The scarlet runner beans he planted that year were very dense and he had tree collards with stalks 2 inches wide. The only thing Panero said he did was continued watering, but, because he constructed a greenhouse and placed his plant life inside, he saw an instant improvement

After the success of his first constructed greenhouse, Panero decided it was time to build another one. This one, however, was much different from the first. "I used all recycled wood for the door on my green house," he said. "We are ripping apart our current deck (all cedar, came with house) and re-milling it with a planner, and we plan on using that wood to frame the green house." He has also been researching home remodeling as an example for building a greenhouse, as he has collected 400 pounds of windows, which he will use to frame his new greenhouse. Along with that, he wants to add a few entertaining additions such as a hot tub in the corner to make it even warmer in winter. "Also, I plan to dig out an aqua culture zone (1/4 of the entire footprint) and grow fish, circulating the fishy manure water through growing beds — increases yield of fish and plants," said Panero. The hoop style greenhouse has worked well, he said, but because they are in this particular house long-term, he wanted to add something a little more permanent and fun.

As Panero has built a few greenhouses in his time, he has discovered there are some advantages and disadvantages to owning a hoop-style greenhouse. Some of the advantages include the temperature. "Early season starts — later season extension," he said. "Tomatoes, and other 'hot plants' love it and go nuts," he said. There are so many greens, less watering that must be done, and less evaporation.

Along with the upside of owning a greenhouse, Panero notes the disadvantages, which he sums up in one word — none. "Who would not want a warm place to grown food longer?" he said. "Hoop greenhouses might not look as pretty as other greenhouses, but they are forgiving, inexpensive, and simple to raise ... no need to be a carpenter." And if you are not the type to build a greenhouse of your own, purchasing a good cloth is a great idea. Panero has a word of caution for beginners about rain, snow, and wind — with a hoop greenhouse, use more clamps than you think, he said. You need to utilize vertical two-by-four supports under the spine. If the smallest amount of rain (or snow) collects on top and the plastic cloth starts to sag, you can say goodbye to your greenhouse and those plants that have had their season extended. "I had a greenhouse collapse and suffocate a couple of beds a year ago due to snow and rain," said Panero. The main reason that this happened was because he had no two-by-four supports and did not have a tight enough spread of the greenhouse cloth. Panero said if he had addressed both of these issues ahead of time, he would have been harvesting through January. "And once I flattened and pulled the greenhouse plastic tight, without any sags or pockets, the hoop popped right back up," he said. "But the plants were dead — cold temperatures, and a blanket of plastic crushed them."

Space Requirements

Consider how much actual space your greenhouse — and you — will require. When estimating this, it is important to take into account the size of your property, as well as the topography of the land. The ground should be fairly level and free of roots and tree branches that could damage the greenhouse.

Because the bulk of the expense will come from the cost of operating the greenhouse, the larger it is, the more expensive it will be to heat, especially during a cold winter. Also, there must be adequate room inside for you to work comfortably and have enough space to store tools and equipment, such as hoses, sinks, pots, watering cans, benches, and accommodate a pathway to move through your greenhouse. If you are skilled enough, you can work from a blueprint to plan the spaces inside your greenhouse. Many hobby greenhouses feature an aisle layout in which two rows of benches are positioned along each side of the greenhouse with an aisle in between. Ideally, benches are oriented north to south. This allows even exposure to the sun as it moves east to west. Many greenhouse gardeners place a small sink and potting area on the north side of the greenhouse.

Space considerations for an attached greenhouse

There are different options and considerations to think about if a greenhouse is going to be built against the side of a house, but generally, the size of an attached greenhouse must fit in with, and be adequately sheltered by the home to which it is attached. For example, an owner might want to build an attached greenhouse with a Gothic curved roof. That could be a problem if his or her house has a low roof **eave**, which is the edge of a roof that projects past the side of the building. This eave will be overshadowed by the roof of the greenhouse. In this instance, it would be in the best interest of the owner to build an attached greenhouse with a lower roof profile that does not obstruct the roof on the main house.

There are other factors to consider as well. For example, how long and wide should a greenhouse be? If you are serious about developing your skills as a greenhouse gardener, you may decide that you will need more space to accommodate your growing skill level and the numerous plants you intend to cultivate. The width of the greenhouse is also critical. Obviously, it is much more comfortable to work in a wide space than a narrow one. A good rule of thumb for the width of a greenhouse is at least 8 feet (2.4 meters). This width allows for a minimum door width of about 22 inches (55.8 centimeters), which most people can fit through, and will also accommodate a 2-foot (0.60 meters) wide walkway and a 3-foot (0.91 meters) wide bench along each side of the greenhouse.

The length of a greenhouse provides a gardener with a little more latitude. It is entirely up to his or her needs. The majority of manufacturers produce greenhouse lengths in standardized 2-foot (0.60 meters) to 4-foot (1.2 meters) increments. These measured lengths are suited to match the size of most common glazing materials found on the market. One of the best ways to determine how large your growing space needs to be is to first decide how many plant trays you plan to place on benches and then use that number to decide the length. For example, it is common for seedling trays to measure 11 inches (27.9 centimeters) by 21 inches (53.3 centimeters). If the benches you will use are 21 inches or 42 inches (106.6 centimeters) wide, that means one or two trays of seedlings will fit on the benches when they are set either lengthwise or width-wise. Bench length can be determined in multiples of 11 or 21 inches. These are standardized lengths.

Here is another way to think about space requirements: Plan to build a greenhouse that is one size larger than you originally intended. That way, as your skills develop, you will have already provided yourself with enough space to accommodate your growing skills and expand the types or number of plants you want to grow. If you build a greenhouse that is too big, you could find yourself with a building that demands more time and energy than you are prepared to give it. However, if the greenhouse is too small, you could crowd yourself out of the space entirely.

The time you intend to devote to your greenhouse is another important issue to take into consideration. Most gardeners know that a greenhouse can require considerable maintenance and work, and though it is certainly not backbreaking labor, it still can be classified as work. No one knows better than a gardener how much nurturing plants need in order to thrive. So, before the construction begins, it is wise to think about the time and energy you are prepared to devote to maintaining your greenhouse and the plants that will grow inside it.

Space considerations for a freestanding greenhouse

As you learned earlier, a freestanding greenhouse can be as large as its location allows. Because it is an independent building, the owner has more freedom to determine the size and shape based on his or her needs and budget. Also, some of the routine tasks a gardener typically must perform, such as re-potting large plants, are easier to do in a larger space. If you plan to hang plants, or place them on shelves, a freestanding greenhouse can provide a tall space with plenty of sunlight that plants will require. Do not forget to leave enough headroom for yourself. Most importantly, a freestanding greenhouse should provide plenty of space in which you can work comfortably.

The height of the greenhouse is important, too, especially in areas where walking is required. Remember the A-frame greenhouse you learned about in Chapter 1? That offered plenty of headroom along the center aisle. If the ceiling is less than 6 feet in height, however, it could be a problem for a taller person who simply wants to walk down the center of the greenhouse. A lower roof could make the space uncomfortable and limit the amount of time a gardener would want to spend inside. Having to continuously stoop while you garden is not only uncomfortable, it can be painful.

A good rule of thumb is to make the height above walkways at least 6 feet, 6 inches in height. The roof height at the sidewalls, however, can be lower — between 4 and 5 feet works well.

A freestanding greenhouse often comprises several different areas that are used by a gardener. There are beds for planting, as well as paths for walking. Some greenhouses are outfitted with quarantine areas in which new plants can temporarily be housed. This can be critical for the overall well being of the other plants, because when a new plant is introduced into a greenhouse, it could possibly infect the space with disease or insects. A quarantine area typically can accommodate two or three plants.

Benches are another common sight in most greenhouses. A **bench** is a portable work surface that is used to hold pots and trays of seeds. Benches can be arranged in various configurations to allow a gardener to perform work in a compact area.

Another crucial space consideration is access into the greenhouse. The path outside the greenhouse should be level and smooth enough, with no sharp turns to allow you to maneuver a wheelbarrow along it. You also should know where the entrance itself will be located. The entrance should allow you to not only reach the garden inside, but also easily push a wheelbarrow through. Typically, the door is located on one end and reaches up to the highest part of the roof. This added height allows you to enter and exit the greenhouse without having to duck down every time to avoid hitting your head on the frame.

One of the main advantages of a freestanding greenhouse is that it can be located anywhere on a property. It can be constructed with its own design aesthetic, independent of the main dwelling. Also, a freestanding green-house can be as large or as small as the size of the property allows. This freedom provides an opportunity to plan the layout of the building with the optimum amount of working space and storage. Depending on the size of the greenhouse, there could be room for tool cabinets, pegboards, and even a shed to store larger tools and equipment.

However, it is important to consider the size of the property on which a freestanding greenhouse will be located. How much of the property will the greenhouse actually use? Again, because it is an independent building, the more space the greenhouse requires, the more expensive it will be to heat during the winter, which will mean higher heating bills.

The site that is chosen for the greenhouse also must be free of trees, hills, or other obstructions to the south, southeast, and southwest. Most plants and flowers are **heliotropic**, which means they track the sun as it moves from east to west across the sky. A freestanding greenhouse must have adequate exposure to sunlight, and it will take up land that might have been used for potential garden beds.

Now that you are familiar with the issues you need to think about when choosing a greenhouse, take another look around your property. Chances are you now have a better idea about your gardening needs and where to locate your greenhouse so that you can fulfill your dream. You have been able to find level ground, where the building can be sheltered by a tall hedge or stand of trees. Perhaps you can already imagine the exact structure and its interior, with the benches laid out the way you want them. If you have that image in your mind, it is time to start making it a reality.

Summary

When you are choosing a site for a greenhouse, it may seem like there are so many considerations to think of. However, the most important thing is to just remember that where you place the greenhouse is all about what you want, not what others want. If you feel that you want a greenhouse close to your home, or attached to it, then that is your choice. You can have a large greenhouse, or a small one, depending on your needs and the space required.

The most important thing to remember when you are putting your greenhouse location plans down is that you choose a place that has:

1. Flat ground that is easy to build upon.

2. An area that is not shaded and gets at least six hours of sunlight.

3. Enough space for you to work in comfortably and grow what you want.

This is an Orangery Greenhouse located at the Palace in Austria. Orangeries started out as buildings with large south facing windows used to collect the warmth of the sun to conserve citrus trees during the harsh winter months. Today they are constructed similar to and function the same as a conservatory or greenhouse.

Chapter 3

Planning Your Greenhouse

> *"My hobby is gardening, I love it, it's my main hobby. I like being at home and I'm very happy being in my house, I love cooking."*
>
> — Susan Hampshire, English actress

I t is the most important part of the greenhouse construction: the planning stage. Keep in mind the old axiom: "a failure to plan is a plan for failure." Your plan will have to take into consideration not only your needs, but also your budget. You must select the proper materials, and if you are going to use wood, the most appropriate type. The next consideration is choosing the right tools for the various tasks that will need to be performed. Also, heating, plumbing, ventilation, and electrical needs must be taken into consideration. A greenhouse requires some, or all, of these components if it is to operate at maximum efficiency. In addition, because building your own greenhouse is going to involve the use of tools and perhaps electrically powered equipment, you should be familiar with some basic safety precautions to help you avoid accidents and injuries. Finally, most areas of the country require building permits to erect structures, such as greenhouses.

Building Materials

A greenhouse can be made from a variety of materials. Some gardeners like a structure made with galvanized steel but others prefer PVC pipe. Many gardeners choose to build their greenhouses with wood, because it is the easiest to work with and relatively inexpensive. Steel is a bit harder to work with and special tools will be needed; it is also heavier and will need greater structural support.

Here is an example of a steel base. Photo courtesy of Juliana America, LLC.

Steel and PVC pipe

Galvanized steel for framing a greenhouse has definite advantages. When steel is galvanized, it has been coated with zinc, which helps to prevent corrosion. Steel is very strong and durable.

PVC or polyvinyl chloride is a tough, synthetic resin pipe, making it a good choice for beginning greenhouse gardeners. However, because it is lighter in weight than steel, PVC can be more easily damaged by high winds. PVC is better suited to small greenhouses.

Types of Wood

There are several types of wood that are favored by greenhouse owners, because many of them are durable against severe weather and aesthetically pleasing to the eye.

- **Redwood**, for example, is tough and naturally resists decay.

- **Cedar** is a high-quality wood, characterized by a pleasant, fresh smell and is red in color.

- **Pressure-treated wood** is another popular choice because it is rot-resistant, which is helpful given the humidity that is typically present in a greenhouse.

It is important to note that wood framing, unlike aluminum or galvanized-steel framing, often requires regular maintenance, because it is heavier and more difficult to work with.

Choosing The Right Tools for the Job

Now that you are familiar with the materials that are used in greenhouse construction, it is time to learn about the tools you will need to put the structure together. Regardless of whether you decide to build an attached or freestanding greenhouse, and do-it-yourself or use a pre-fabricated model from a kit, you will require certain tools to get the job done properly.

The following tools can be used to build a greenhouse from a kit:

- **A variety of wrenches** - A monkey wrench would be best because you can change its size to fit the various bolts you may have to attach for the greenhouse.

- **Pliers** - Needle-nose pliers would be best.

- **Vice grips** - This will help you keep things secure while you are attaching panels, walls, and more.

- **Portable, electric drill with bits** - If you have to look around for a place to plug in your drill, you may not be able to build your greenhouse where you want.

- **Hand saw or hacksaw** - Handheld tools are the best and you will only have to do a bit of cutting.

- **A ladder** - You will need to attach a roof to the tops of walls and a ladder will be needed.

- **Work gloves** - After your first sliver, you will realize how important work gloves are.

- **Protective eyewear** - All it takes is one piece of plastic, wood, or steel shooting off while you are working on the greenhouse and landing in your eye to show the importance of eyewear.

- **Ear plugs** - If you are cutting anything with a powered saw, hammering nails in, or using a portable drill, you will want to protect your hearing.

To build a conventional or do-it-yourself greenhouse, the following tools will be required:

- **Tape measure** - Always remember to measure twice and cut once.

- **Hammer** - If you are building with wood, this is essential.

- **Screwdrivers** - Screws tend to be better at holding pieces of your greenhouse together than nails, so you will need a screwdriver.

- **Wrenches** - Attaching brackets will require wrenches for the bolts.

- **Saws** - If you are cutting plastic or wood, you need saws.

- **Plane** - You may have to shave off a small part of the wood you are using to build the greenhouse and the plane will do that.

- **Framing square** - Angles are very important when building a greenhouse.

- **Level** - Your walls need to be level, and a level will help you determine if they are.

- **Chisels** - If you have to make notches for hinges, chisels are very important.

- **Mallets** - Securing glass into a spot, putting a door in, or if you just need to hammer two pieces of wood together without nails, use a mallet.

- **Portable, electrically powered drill and bits** - Same reasons as cited in the last list.

- **Caulking gun** - If you are working with plastic and/or window, then a caulking gun is important to ensure that you do not have heat escaping through the window.

Make sure you know how to use the required tools and ensure that they are all in good working condition. On any construction project, large or small, working safely should always be an important consideration.

Accidents happen fast, usually in the blink of an eye, and often when someone is momentarily distracted. Ironically, most accidents are easily preventable.

For example, many people believe that they know how to do something as seemingly simple as use a ladder. But, too often that is not the case. The following lists contain safety tips for each stage of your greenhouse building to avoid accidents and injuries. It is a good idea to carefully inspect the ladder before you use it. Always make sure the ladder is in good condition, and that you are using the ladder properly by checking the following conditions:

- There are no cracks or corrosion.

- The ladder's feet are planted on stable, even ground.

- The ladder's base is planted 1 foot away from the building (of an attached greenhouse) for every 4 feet in eave height. Thus, if the eave is 16 feet high, the ladder's feet should be planted 4 feet away from the base of the house.

- There is nothing on the ladder that could cause you to fall or lose your balance (including things that you are carrying).

Another good safety precaution to follow is to check that tools you will use are in good working condition. Make sure your hammer meets the following requirements in order for optimum usage:

- Is the head firmly in place? If it is not, it could fly off and hit someone or break something in front of you. If the hammer head is not in place properly, buy a new hammer or use wood glue to secure it.

- What about that saw? If the saw is dull, it will not cut things in the amount of time you want it to.

- Is the handle securely attached to the blade? If the handle is not securely attached to the saw blade, you could end up hurting yourself when the saw comes off the blade and you move forward while the saw stays where it is in the wood.

If you plan to use tools and equipment that is electrically powered, ensure they are in proper operating condition. To inspect them safely, disconnect them from the power source. This will bring them to a zero-energy state. A zero-energy state means that energy powering the equipment, such as electricity, air, or hydraulics has been released. This will prevent an unexpected start-up.

Another way to stay safe as you work is to wear appropriate personal protective equipment, or PPE.

- Wear proper eye protection. Safety glasses will protect your eyes from flying debris.

- A good pair of work gloves will protect your hands from splinters, when you handle wood, and cuts, from sharp objects and tools.

- Wear sturdy work boots with steel toes and thick soles. They will protect your feet if a heavy object falls on them, or if you step on a sharp object like an upright nail head.

- You should wear a helmet. Things can fall on you, especially when you are working with other people. Head injuries are a common workplace accident in construction and you need to protect your brain.

- Ear protection is important because you will be working with power tools and long-term exposure can result in hearing loss.

Planning for Climate Control

As you already know, plants require consistent environmental conditions. Heating, cooling, and ventilation play a critical role in maintaining a nurturing environment. To ensure a greenhouse maintains a plant-friendly environment, you may need to install a number of systems including heating, cooling, ventilation, misting, and lighting.

- The perfect growing temperature for plants is between 50 degrees Fahrenheit (10 degrees Celsius) and 85 degrees F (29.4 C). It is best to keep your plants in this range. If the temperature falls below 50 degrees, the plants will die, and if the temperature goes above 85 degrees, the plants will experience slower growth and flower drop that prevents the plants from producing fruit.

- Humidity is important in the greenhouse. Typically the humidity should be between 45 and 60 percent. If it is above 80 percent, mold can grow on the plants and in the greenhouse, and if the humidity is less than 45 percent, the air can be too dry for the plants.

This is especially challenging in climates that experience cold winters with heavy snowfall. Supplemental heating is typically required to keep plants healthy. There are a variety of methods to heat a greenhouse. For a freestanding structure located in a mild climate, an electric heater can be a good choice. Some owners install infrared heaters in the ridges of their greenhouses. This technology radiates heat that is similar to the warmth of the sun. Another method for heating a greenhouse is a kerosene heater. However, this type of heater can cause injury to plants by causing them to wilt and lose their green color over a long period of time because of the fumes released by the combustion.

The amount of heat needed to warm a greenhouse depends on three things: the size of the space, the type of covering material, and the climate in which the greenhouse is located. For an attached greenhouse, connecting to the home heating system might make the most sense. Fuel can be expensive, and tying in to a main system could help cut down on heating costs during the winter.

Adequate heating is a critical consideration. To better gauge how much heat a greenhouse will require, it will help to understand how heat is actually measured.

Heat is measured in **BTU or Btu**, a traditional calculation that stands for British thermal units. That equals the amount of heat needed to raise 1 pound of water 1 degrees. To calculate the amount of heat required to heat a greenhouse, it is necessary to figure how many Btu of heat output will be needed. First, multiply the total square footage of the greenhouse panels times the difference between the coldest night time temperature and the minimum night time temperature the plants will require. Now multiply that figure by 1.1. The traditional formula will look like this:

$$A \times D \times 1.1 = BTU's$$

For example, take a greenhouse area of 3,800 square feet and 45 degrees as the difference between the coldest night time temperature (10 degrees) and the preferred temperature (55 degrees). So, 380 square feet x 45 degrees x 1.1 equals 18,810 Btu.

If a greenhouse based on the size of this example has insulation or double-glazed glass, you can subtract 30 percent from the total Btu to account for the building's ability to better retain heat. An attached greenhouse with the same dimensions would require 60 percent fewer Btu, because it is attached to the wall of a house and benefits from the added insulation.

Many greenhouse owners place smaller heaters inside the buildings. These devices usually have a thermostat switch so it can be turned on whenever necessary. But there is a problem — plants do not like noxious fumes such as those produced by kerosene heaters, because they are toxic to the plants. If kerosene has sulphur in it, the sulphur fumes can damage the plants because they are not able to turn it into oxygen. Except for electric heaters, most heating devices release fumes that can severely damage the growth of plants or destroy them. Gas heaters require venting to ensure fumes do not affect the plants.

Another problem is that without effective air circulation, the plants located nearest to the heater can overheat. Likewise, the plants furthest away may freeze. Remember, plants and flowers require consistent climactic conditions. One solution to help avoid damaging plants from heating devices is to use electric heat. There are no dangerous fumes. A thermostat, an electronic device that can be programmed to switch a heater on and off, can control electric heaters, such as radiators, to help maintain stable temperatures. An alarm system can also be installed to signal if the temperature drops to a dangerous level. Several thermometers should be placed at bench

level at different locations in the greenhouse so you can see whether or not heat is being distributed evenly. A plant's health depends on consistent environmental conditions. Most heaters have an automatic on and off switch. Most kerosene heaters do not have a thermostat, which means control over

Here are examples of different thermometers that can be used in your greenhouse.
Photo courtesy of Juliana America, LLC.

the environment is limited. As is the case with most homes, mechanical thermostats are commonly used to control heating and cooling systems. If a gas heater will be used, it is critical to move the combustion that creates the fumes out of the greenhouse. It is quite common for an oil furnace that heats a greenhouse to be located in a separate building.

For an attached greenhouse, the heat source is generally the main house itself. Doors and vents located between the greenhouse and the house allow both warm and cool air to pass from one room into the other. In effect, it is like adding a bedroom or den to the main house and the costs normally associated with that.

A Heatmat Thermostat ™ is a seedling heat mat. It provides an even-temperature to encourage seeds to germinate.
Photo courtesy of Juliana America, LLC.

Heating and Insulation

Although the sun is the primary source of heat for a greenhouse, most greenhouses need additional heat to maintain a comfortable, stable environment for the plants inside. Most plants do not thrive in an environ-

Bubble insulation is an easy way to lengthen your greenhouse season. It can be used to insulate your entire greenhouse or specific areas.
Photo courtesy of Juliana America, LLC.

ment that has extreme temperatures and extreme moisture. This is why heating, cooling, and ventilation are crucial to maintain proper temperature and humidity.

The basic science works like this — plants will thrive when light and moisture, heat and humidity, and fertilizer are all available in adequate supply. All plants derive their energy to grow from carbon dioxide and water absorbed through their roots and leaves. When exposed to light, these molecules convert carbon dioxide into compounds such as sugar, which is the foundation of the plants' ability to grow — a process known as **photosynthesis**. During this process, plants release oxygen, heat, and water. When a gardener can control the air temperature and quality, as well as exposure to light, the plants inside the greenhouse will grow.

This is an example of a FlexiFurnace™ heater.
It features automatic and manual
temperature controls.
Photo courtesy of Juliana America, LLC.

Typically, a greenhouse environment is cool, warm, or hot, depending on the types of plant life it is intended to support. A cool greenhouse has a minimum night-time temperature of 41 degrees (5 C) and is used to start seeds and support cuttings early in the year that will be planted in early summer. A cool greenhouse environment is also appropriate for vegetables and many robust plants.

Greenhouse Heating

Heat naturally flows from a warm space to a cooler one. In warm weather, heat moves from the outside of a building to the inside. In colder weather, the opposite is true — when the temperature drops, heat will flow from a warm interior to a cold exterior. In winter, the heat from a greenhouse that is lost must be replaced by whatever heating system is in place.

As you have already seen, heating a greenhouse during the winter can be expensive in a colder climate. The heat lost through glazing can be ten times as much as that lost through regular insulation found in a house. However, there are materials and methods that can help curb heat loss. For example, double and triple glazing is one method for reducing heat loss. Double-glazing can be as much as 30 percent more efficient at reducing heat loss than single glazing. Using triple glazing can cut heat loss by half as much again.

Glazing refers to transparent glass or plastic. Glazing should let in the maximum amount of light and deter heat loss. Traditionally, glass is the material used for greenhouse glazing. However, polycarbonate glazing is also popular because it is light, strong, and shatter-resistant.

Insulation as an Alternative

Installing insulation in the greenhouse is another solution. Not all greenhouses are built completely out of glass or plastic. There are many that are built with sides filled with insulation. Quite often, greenhouse owners insulate both the sidewalls and endwalls of their buildings to help reduce heat loss through the walls. The insulation is installed up to the height of the bench. Some common insulating materials include polystyrene and polyurethane insulation boards. Another advantage in using insulation is that it easily can be attached to the greenhouse frame for the winter months and then just as easily removed for the spring and summer.

Insulating walls and floors provides resistance to the flow of escaping heat.

Insulation materials such as polyicynene, polystyrene, and polyurethane are effective because they function by limiting the movement of air. Reflective insulation reduces the amount of energy that travels in the form of radiation. **Batts, loose-fill, blankets,** and foam are the most effective insulation materials, which explains why they are used so often to insulate residential buildings. A blanket is made of batts or rolls of fiberglass insulation.

An alternative method of insulating a greenhouse is the use of a thermal blanket. A thermal blanket is similar to a shade and is installed on a track or cable and drawn over the glazed areas of the greenhouse for the night. In the morning, the thermal blanket can be drawn back to expose the plants to sunlight.

Still another technique that some gardeners use to help insulate their greenhouses is to utilize the landscape around the structure. Cultivating low-growing plants, such as shrubs around the foundation, or planting a hedge that acts as a windbreak can help insulate the greenhouse from cold, harsh winds.

There also are some other, relatively simple methods of saving energy in a greenhouse. Just adding a layer of a clear plastic sheet between the inside of the greenhouse and the insulation layer provides a space of static air between the permanent cover and the insulation. This will help further insulate the building. If the covering on the greenhouse is dirty, for example, that will reduce the amount of light that is able to enter the greenhouse, thereby reducing the amount of heat inside. Simply keeping the covering clean will help retain heat inside the greenhouse. Also, watering the plants while the sun is high allows the wet surfaces to dry out and warm up before dark. Concrete floors, walls, and pathways absorb heat. At night, that warmth will be released through the windows, walls, and anywhere that heat can escape.

Ventilation

Consistent temperatures are crucial to the health of plant life. But there is another component that works in partnership with light, humidity, soil, and warmth: oxygen. Simply put, plants need air. And, because plants also breathe in carbon dioxide and release oxygen, a greenhouse must be able to provide a plentiful supply of fresh air, which is why proper **ventilation** — the movement or circulation of fresh air through a space — plays a critical role in the success of your greenhouse.

Typically, fans and vents are used to ventilate a greenhouse, forcing out any hot air, reducing the humidity, and circulating fresh air. In a greenhouse, one of the main challenges is to provide a source of fresh air that is neither too hot nor too cold. Thus, a ventilation system must be able to take in outdoor air and expel indoor air so that the temperature inside the greenhouse does not become too cool, as this will cause the plants to wilt.

It is quite common for a freestanding greenhouse to have a fan located at one of the ends of the building and a vent at the other end. Cool air close to the floor gets heated and rises. As the warm air naturally escapes out of the roof vents, it is replaced by cool air that enters through the side vents. Typically, a vent is hinged on one side and can be easily opened with levers. The fan is positioned as high up in the greenhouse as possible so that it mixes the cold air with the rising warm air, balancing the building's interior temperature.

For an attached greenhouse, the house to which it is secured can often provide the required ventilation. It is a big advantage, because this method of ventilation provides fresh air that is already the correct temperature. If the greenhouse also has a sliding glass door or window that opens into the house, that is all the ventilation the greenhouse will need. Air will naturally flow from the main dwelling into the greenhouse. If the attached greenhouse also has a door or window that opens to the outside, that opening would provide adequate ventilation in warmer weather. Remember, as cool air near the floor is heated, it becomes lighter, rises, and escapes through the roof vents. That warm air is replaced by cool air that enters through the side vents. Regardless of the type of greenhouse you decide to build, the plants inside will require consistent environmental conditions, and adequate ventilation is a key component for their health. Too much carbon dioxide can kill the plants through suffocation.

Below are examples of Ventilation Systems. Photos courtesy of Juliana American, LLC.

Here is an example of a shutter vent system located on the side of the greenhouse.

This is an example of a fan located at the end of the greenhouse.

A solar vent is a solar-powered fan that circulates air through your greenhouse.

Plumbing and Electricity

Generally, most types of greenhouses are going to require electricity and plumbing. However, there are exceptions. A sunroom, for example, will probably not require electricity and plumbing, because the roof itself is really part of the main house and can draw both electricity and water from it. A small, simple freestanding greenhouse, located in a temperate climate zone, probably will not require electricity because warmer air and humidity are available year round.

It takes a certain level of skill to properly and safely connect utilities, and not everyone is comfortable with the task. In many instances, it is recommended to leave those tasks to a building professional, and in some areas of the country, building codes may actually prohibit anyone other than a licensed electrician or utility professional from performing these tasks. Because the primary purpose of a greenhouse is to maintain consistent environmental conditions for its plant life, providing the necessary power and plumbing is critical.

Water Lines

A water line is a pipe that moves water from one location to another and is usually installed while the foundation of the greenhouse is under construction. The pipe should be laid below the frost line. A backflow prevention line also should be installed to prevent the contamination of the main house's drinking water from exposure to used water from the greenhouse. The pipe is connected to a dry hydrant in the greenhouse. If the temperature drops below freezing during the winter, the hydrant can be drained to prevent the water pipe from bursting. Because many greenhouses have sinks, the drain pipe should be connected to a sewer line or septic system to allow adequate drainage. Connecting water lines to a building is not always a simple job. Depending on your experience, and the local building codes, you might want to consult with a professional, or hire one to do the work for you.

If the greenhouse is located close to the main house, the job of providing water is not complicated. Simply run a hose from the outdoor tap on the side or back of the home to the greenhouse. However, depending on how severe the winter is where your greenhouse is located, this may not work because the water could freeze in the hose. If this is the case, a permanent water line, as described in the previous paragraph, will need to be installed.

The black box in this picture is a drip irrigation and fertilizer system. It provides the plants in your greenhouse with an adjustable amount of water and fertilizer.
Photo courtesy of Juliana America, LLC.

Electrical Work

As with proximity to a water supply, a freestanding greenhouse that is located close to a house will make it more convenient to provide electricity. An easy solution is to run a power cord from the greenhouse to the **junction box** in the house, usually located in the basement. A junction box is a container that houses and protects electrical circuits. To avoid potential safety hazards such as tripping in the garden or cutting the power cord and getting an electric shock, you can set and run the cord along the boundaries of the garden rather than across it.

Electrical safety precautions

More often than not, a freestanding greenhouse will be connected for electrical power by an underground cable. Again, installing underground cable is not a task that everyone should perform. A mistake involving electricity can be dangerous — even fatal. And, depending on where you live, the local municipal codes may require you to hire a professional electrician to perform the work. It is recommended that you check with the local utility company before starting any installation to avoid hitting power and water lines. If an electrical cable will be installed, then it must be installed at least 2 feet (61 cm) below the ground, or according to the local building code. Although this is a general measurement for greenhouses, the depth can vary depending on your local building codes. However, when it comes to working with electrical cables, it is safer to err on the side of caution and consult safety manuals, or hire professionals.

Everyone knows that water and electricity do not mix, which presents a unique problem within a damp, humid interior of a greenhouse. Because of this humidity, there also are precautions to take to prevent accidents, such as electrical shocks and burns. Electrical switches and fixtures should be installed in waterproof outdoor-rated housing. All cables between the junction box in the house and the greenhouse should be waterproof. Outlets also should be waterproof and installed higher up the walls of the greenhouse; that way, the outlets are less likely to get wet and cause an electrical shock. Remember, electricity and water do not mix — always use waterproofed outlets to help avoid being electrocuted.

Because there will — no doubt — always be water in the greenhouse, ensure all electrical circuits are connected to a ground fault circuit interrupter, also known as a GFCI. A GFCI is a critical piece of safety equipment on a construction project whenever there is a potential electrical hazard, and it can ultimately save your life. Basically, a GFCI is a sensitive circuit breaker that compares the amount of current traveling to and from electrical equipment. If the current traveling from an electrical source is different from the current returning to it, the GFCI trips the circuit, preventing electrical shock. If it did not, a person using electrical equipment, such as an electrically-powered drill with an exposed wire on its cord, could receive an electrical shock if he or she came in contact with the exposed wire.

An electrical shock or burn occurs when a part of a person's body completes a circuit, connecting a power source, such as an electrical line with the ground. For example, suppose a person is operating a defective, electrically powered saw that was plugged into a generator with a GFCI. The GFCI would sense if there was a defect in the saw and trip the circuit, preventing an electric shock.

If a generator does not have a built-in GFCI, use a portable GFCI, which should be plugged into the power source. Then, an extension cord can be plugged directly into the portable GFCI. When the tools are plugged into the extension cord, there will be less of a chance of an accident.

Because working safely with electrical equipment and cables requires experience, it is recommended that you seek the advice of a professional to determine that all connections are safe and power sources meet the regulations of local building codes. Depending on the municipality, electrical work, such as installing power lines, might not be considered a part of the construction work. The municipality, such as the town you live in, the city, or even the county and state where you reside, can require a permit for any installations or alterations of electrical equipment. Other projects, like installing plumbing, could be regulated under a separate municipal code and require a different permit. Contact your local city hall or municipal building and ask for the required electrical codes, as well as the building codes for your area. You can also talk to building inspectors to find out what would be a pass or a fail with a building inspection.

Wiring correctly

Generally, for a small greenhouse, less than 500 square feet (152.4 meters), a single 20 amp, 120 volt line will provide enough power for a fan, louver, or vent motor, room light, and control switch for a gas heater. When planning for a large greenhouse, figure out the wattage of each piece of electrical equipment, and simply add these together to determine the total electrical need for the greenhouse.

Installing Receptacles

Inside the greenhouse, the electrical cable is attached to a distribution box that provides power to the various appliances. The distribution box should have a main switch, as well as individual switches for each circuit. Plastic conduit, which houses electrical wires, is recommended for installing and attaching wiring to each appliance. Waterproof outlets and switch boxes also are recommended because of the humidity inside the greenhouse. If frequent power outages are common in your area, it is a good idea to use a backup power supply, such as a generator. If there is a power outage during a cold spell, the last thing you want is for your plants to perish because of plummeting temperatures.

Will You Require a Building Permit?

As you get closer to the actual start of your greenhouse construction, there are still some important issues you need to address before you begin to actually build your greenhouse. First of all, find out whether or not you will require a building permit. Building codes and planning regulations are set by municipal planning departments and typically regulate electrical, plumbing, construction, and permitted square footage for structures. These regulations can definitely have an impact on the size of the greenhouse.

Zoning regulations also will play a role choosing a location on a property where a greenhouse is to be constructed. Again, check with your local municipal government office or chamber of commerce for more information about zoning regulations. Many of these regulations can be accessed online. Check your local municipal government offices for details. Or, call your local city hall or town hall clerk. They can direct you to more information.

Determining the Need for a Building Permit

Typically, when a greenhouse is constructed on a permanent foundation, a permit is required, because construction can involve using utility lines.

Another issue is the type of glazing materials that have been approved for use in buildings. For example, in areas where hurricanes are common, some municipalities require all buildings to use hurricane-proof glass. If it is necessary to apply for a permit, a plan of the greenhouse will be required so that building officials can see that it will be built to the current codes.

Building, plumbing, and wiring codes exist in nearly every city or town and should be followed. All codes contain minimum requirements for materials and procedures. If you are having your greenhouse built by a professional, or are hiring a professional for the plumbing or wiring of your greenhouse, you should ensure that the person you hire is licensed and bonded, and always follow the local building codes.

Now that you have become familiar with how to plan your greenhouse, you are well on your way to beginning its actual construction. Whether you are planning to build a do-it-yourself attached structure, or construct a free-standing hobby greenhouse from a kit, you have the basic knowledge you need to get started. You also are familiar with the basic safety precautions to follow to ensure you get the job done without injury, so you are ready to take the next step in getting in touch with your natural side: construction.

Summary

It might seem like plenty to assimilate when you have to think about water drainage, ventilation, building permits, utilities and more, but the more complicated a greenhouse you build; the more you will have to deal with these things. You will have to deal with electrical, plumbing, heating, and more. If you have a lean-to greenhouse, you may not need to deal with that, but with A-frame houses and larger houses, you need to consider these things. You also have to make sure you have the right tools for the job and all of this is not to complicate things or worry you about everything you need to do; it is to make things easier on you. In addition, with building permits and everything associated with utilities, these exist to keep you safe. Building a greenhouse is building a building that you will work in, and it has to be built correctly in order to keep you — and anyone else who enters — safe.

Chapter 4

Building Foundations

> "There is no gardening without humility. Nature is constantly sending even its oldest scholars to the bottom of the class for some egregious blunder."
>
> — Alfred Austin, English poet

The amount of time it takes to build a greenhouse depends on three things — the type of greenhouse (attached or freestanding), how sophisticated you want it to be, and your budget. For instance, a simple 5 x 5-foot pop-up freestanding greenhouse from your local gardening center can be assembled in a matter of hours. On the other hand, if you plan on a large greenhouse with a concrete foundation, that can take up to a week. Likewise, an attached greenhouse, much like an added den or bedroom can take days. The first step, however, is choosing a level surface on which to build.

Choosing a Level Surface

The next stage in building a greenhouse is making sure the site for the building is fairly level. Earlier in this book, you became familiar with how to choose a site for the greenhouse so that it will receive the optimum exposure to sunlight, but that is not the only thing you need to think about when choosing a location. Another important consideration is whether the greenhouse will be constructed on ground that is level or that can be made level with a minimal difficulty. For example, a location that slopes down slightly to the south will work well for drainage; however, if the ground is really uneven, or is located below a hill or steep-slope, then water will have to be directed away from the greenhouse. Use a **level**, a glass tube that contains liquid with an air bubble inside it, to determine if the site is basically even. It is placed on a surface and measures whether or not the surface is truly even.

The level will indicate when a surface is both level and **plumb**, meaning a vertical surface, such as a wall or door, is perpendicular to a horizontal surface, like a floor. If measurements are uneven, the entire structure will be crooked and could possibly collapse.

1. Mark out the area where you are going to building the greenhouse.

2. At each corner where you are going to place the corners of the walls, put a small two-by-four down.

3. At each two-by-four, put a level down on the board to determine whether or not it is level.

4. If one corner is not level, build it up with dirt to make sure it is level and then check it with the two-by-four and level again.

Testing For Appropriate Sunlight

To determine if you have enough sunlight, on a day in the summer, either watch from your home or go out for the day and determine how much sunlight hits the ground during a given day.

- If on a sunny day no sunshine hits the ground, then the area is completely shaded.

- If on a sunny day three hours of sunshine hits the ground, then it is partially shaded.

- If on a sunny day four to six hours of sunlight hits the ground, then it is partly sunny.

- If on a sunny day at least six hours of sunlight hits the ground, then it is full sun and perfect for your greenhouse.

Preparing the Site

Once you have chosen the actual location, and tested it for appropriate sunlight and level, the first step in preparing the site for greenhouse construction is marking the location of the building's foundation. On the site that you select, measure the dimensions and mark the corners with standard 2-foot high wooden two-by-two stakes. To ensure the perimeter is square, measure the diagonals — they should be equal. Attach string to the top of each stake and pull each piece taut so that the entire perimeter is now marked.

If you plan to install a concrete foundation for the greenhouse, set up **batter boards** at each corner of the perimeter to determine if the ground is level and where the greenhouse will be. A batter board is a horizontal board that is attached to posts located at the corners of the building site that shows the accurate layout of a foundation.

1. Drive three two-by-four stakes 2 feet outside the corner stake.

2. One stake should sit directly on the diagonal across from the opposite corner.

3. The other two stakes should be driven 4 feet from the stake, and positioned down each side of the perimeter.

4. There should now be a 90-degree angle formed.

5. Connect the stakes with one-by-four boards nailed level near the top of the stakes.

6. Make a small cut in each of the batter boards so they correspond with those original lines.

7. Use a level to mark an even elevation and attach the strings at each batter board.

8. Next, remove the original stakes and strings. You now have a square and level frame that matches the greenhouse site. The intersections of the remaining strings mark the corners of the greenhouse.

Constructing the Foundation

Regardless of whether a greenhouse is temporary or permanent, it should not be built on a dirt floor. The earth will eventually become muddy and attract diseases, weeds, and other pests. Again, your budget and how much time you intend to spend gardening inside your greenhouse will help you determine whether you construct a simple, small hobby greenhouse, or a larger, permanent structure with a concrete foundation. A freestanding greenhouse should be built on some type of solid surface. You do not want wet soil as a floor in a greenhouse, because it will help breed parasites that could spread disease to your plants, as well as affect the temperature. There are four common types of foundations for freestanding greenhouses: wood frame, concrete slab, concrete footing and walls, and anchor stakes, which will be discussed later in this chapter.

A firm foundation is a critical component for a greenhouse. If the base of the building is not stable, the whole building could shift, which can damage the frame and windows and could cause damage. If this happens, nothing will fit properly, including doors, windows, and glass. It must be located on a relatively level patch of ground, regardless of the type of foundation it has.

The climate and the materials that will be used to construct the greenhouse play an important part in choosing the most appropriate type of foundation for building. For example, if you live in an area that regularly experiences a very cold, snowy winter, a poured-concrete foundation will be the most durable.

In geographical areas where the winters are not bitterly cold and filled with snow, a portable, freestanding greenhouse, covered with polyethylene, would only require a base and stakes to hold it in place. Without an adequate anchoring system, a storm or strong wind could severely damage or destroy a greenhouse.

A temporary greenhouse can rest on a foundation made of treated lumber. Another low-cost choice for this type of greenhouse is a wooden post foundation. This type of base is made with pressure-treated wooden posts and is a good choice for a greenhouse constructed from wooden frames.

Many greenhouse owners use a poured-concrete footing that is about 4 to 6 inches wide. This is a relatively easy foundation to build and will act as a base for a concrete greenhouse knee wall. The **knee wall**, typically 2 to 3 feet in height, extends from it. In colder climates, a knee wall that extends below the frost line will prevent the concrete slab from cracking or shifting.

The footing can be set below the frost line to support a foundation wall erected on it. Some greenhouses are constructed with 2- or 3-foot high concrete, stone, or brick erected on a footing. Before concrete is poured, however, the electrical conduit, the water line, and the drain line should be installed. These need to be installed first because they will often be coming from under the ground and you do not want to break through your concrete to bring in the water line.

Even if you choose to hire a professional to build your greenhouse, it will help you to understand more about the products that will be used in the construction. Concrete is a common material, especially for freestanding greenhouses. It is made of sand, Portland cement, and a mixture of limestone and clay or marl. Most permanent foundations for greenhouses are made with sand or gravel. A concrete slab is typically 4 to 6 inches thick and poured on a base of gravel to provide drainage. Water will move through gravel and drain. Standard blocks measure 7-5/8 x 15-5/8 x 7-5/8 inches. When a block is set in mortar, it measures 8 x 8 x 16 inches. The blocks can be laid out in a wall in multiples of 8 inches.

A foundation can also be comprised of brick. A brick is made of a mixture of shale and clay, which comes in many styles and colors. Other types of brick can be found in fireplaces and walls. A greenhouse foundation can also incorporate pavers, which are typically used to build walkways. Like bricks, they are durable and weatherproof. Pavers are small stone tiles that are used to make cobble streets in some cities. They are often used in backyards as a walkway towards a fire pit.

The first step in laying out a foundation is to mark the dimensions of the foundation and make sure it is perfectly square. Remember the term plumb? It is the measurement that indicates when a vertical surface is exactly perpendicular to a horizontal surface. Think of a wall where it meets the floor — if the foundation is even slightly out of plumb, the greenhouse will be off-balance, meaning that doors and windows will not fit properly. Use the level to ensure the foundation is squared.

The next step is to determine where the four corners of the foundation will be. To do this you will:

1. Drive a stake to mark the location of one corner.

2. Set one 3-foot length of wood in one direction from the corner.

3. Set down a 4-foot length of wood so that it is perpendicular to the first length of wood.

4. Place a 5-foot length of wood between the other two lengths of wood so that it lies on a diagonal. This will form a triangle that will be square.

5. This gives you one full corner and two half corners.

6. At this point, do the same diagonally across from the full corner to create all four corners. At this point you would have four boards making up where the walls will be, and one 5-foot length of wood running between the two. This will make it look like there are two triangles making up the floor.

Wooden Foundation

One of the most common greenhouse foundations you will find is a wooden one. It is inexpensive and is often the preferred choice for many greenhouse gardeners. A wood frame foundation is one of the easiest to make. It can consist of treated lumber or landscaping timbers, and it is recommended that you use a naturally resistant wood. In Chapter 3, you learned about different types of wood and their characteristics. Woods like cedar and redwood naturally resist damage caused by rot. Pressure-treated wood is another popular choice for greenhouse construction. If you are a do-it-yourself type of person, then you will be able cut the wood to size with a standard hand or power saw. Do not forget to wear your PPE (personal protective equipment), including protective eyewear, which will prevent pieces of wood or other debris from hitting your eyes.

Once you have cut the wood to fit the greenhouse, lay the boards together on the ground the way you are going to set the greenhouse on them. To ensure you build on a solid foundation and have adequate drainage, make sure the site for the foundation is on level ground. Measure, square up, and mark the perimeter.

1. Make sure you remove any rocks, sticks, or dirt clogs that could keep the board from laying flat on the ground.

2. Set your level on top of one of the boards. The horizontal bubble in the level should be floating in between the two lines marked on the level.

3. Perform this measurement for each board to make sure that the greenhouse base is level.

Now that the foundation has been leveled, it is time to secure the timbers together with lag screws. A **lag screw** is heavy with a square head and is specifically designed to secure wood.

1. The lag screw should be 3 inches or longer than the first board it is screwed through to ensure it is properly secured. In the initial stage of your building process, lag screws should be attached at each corner in the same way.

2. When you have completed this task, lay a protective ground cover under the base and cut away any extra fabric from around the edges. The ground cover should be designed for landscaping, as it will permit water to drain out through it, and, at the same time, prevent weeds from growing up and into the greenhouse.

3. Dig a trench 4 to 6 inches below grade around the perimeter of the base. The width of the trench should be a few inches wider than the lumber.

4. Now, fill the trench with 2 inches of gravel.

5. Stack four-by-fours or timbers on top of each other and attach them with large galvanized deck screws.

6. Double check the timbers to make sure they are level and square.

You can anchor the base to the ground using a few different methods:

1. Drive one-half-inch reinforcing bar, or No. 4 **rebar** stake 3 feet into the ground at each of the corners, as well as placing one every 4 feet along the inside of the foundation. Rebar is a reinforcing bar with ridges for better anchoring.

2. Use Conduit J nails to attach the rebar. Another attachment method is to drill holes through the timbers and drive L-shaped rebar through them into the ground.

3. One more option is to cut two-by-four stakes and secure them with wood screws.

A pressure-treated wooden post foundation can provide a strong, simple base for a wooden frame greenhouse. You may require a posthole digger to dig the holes. Another tool for the job is a **power auger**. An auger is a powered tool used to bore holes in the ground.

Holes should extend down past the frost line. To avoid potential hazards, make sure you contact the local utility companies to confirm the locations of cables and pipes before any digging or construction begins.

Once the holes are dug:

1. Place 2 to 3 inches of gravel in the bottom of each hole, leaving a few extra inches on the height of the post.

2. Cut the post to the right length after it is set in the hole.

3. Backfill the hole and post with dirt, ensuring the post is vertical and squared up with the foundation lines. Again, a solid structure must have a properly constructed foundation to support it.

4. If everything is squared and even, then you have finished building the foundation of your greenhouse, and you are ready to begin the next step: installing the floor and securing the actual greenhouse structure to the base.

There are numerous materials for flooring a greenhouse. Again, the choice of material for building the base really depends on your level of skill, construction experience, and confidence. One of the most popular types of flooring is gravel. Many greenhouse gardeners prefer it, because it is simple to install, relatively inexpensive, and provides reliable drainage. It is also easy to maintain. Follow these steps to install your gravel floor:

1. Clear the area where you intend to place the gravel.

2. Cover it with weed barrier cloth — a thin cloth that you lay down over the ground to keep weeds from coming up. You cut holes in it to allow your plants to grow, but weeds are kept to a minimum. For the foundation, you are putting down the weed barrier cloth without holes so weeds do not start popping up in your greenhouse.

3. Shovel gravel onto the barrier cloth.

Besides gravel, other flooring options include bricks, pavers, and concrete. A concrete floor takes more work to install, but its surface is smooth and easy to clean. A concrete floor also has the advantage of reflecting light and holding in some heat.

The next task is to ensure the base is square. To do this properly, get your tape measure and take two diagonal measurements of the base. One measurement should be from the front left corner to the back right corner. The other measurement should be from the front right corner to the back left corner. The base will need to be adjusted until the two measurements are the same. You are now ready to anchor your foundation to the ground.

Now it is time to secure your greenhouse to the foundation base. If you are building an aluminum frame greenhouse, it is recommended that you use 2-inch galvanized lag screws and washers for this task. A **washer** is a thin plate with a hole in the middle that distributes the load of a threaded fastener. To secure your foundation base:

1. First, drill a small hole in the aluminum ridge at the bottom of the greenhouse frame to begin the attachment.

2. Insert one screw for each panel in the greenhouse.

3. After all the holes have been drilled, place a washer over each hole and secure the screws into the base of the greenhouse.

4. If you want to add extra insulation, you can caulk the bottom of the aluminum ridge sealant at the point where it joins the base. This method will help seal the greenhouse, and during the winter months, it will keep cold air from entering the greenhouse and warm air from escaping the building.

Concrete Slab Foundation

If you are very skilled in construction and comfortable working with cement, then using a concrete slab to build the base of your greenhouse will work for you. A concrete slab makes a solid base for any greenhouse. A concrete slab is the best type of foundation because it is solid, it keeps cold air from the ground rising up, and it prevents weeds from growing. However, a concrete slab is rather permanent and it can be difficult to make if you are not experienced, so sometimes a wood foundation or gravel foundation is a better option. For an attached structure, the finished floor is typically set level one or two steps below the floor of the main house.

For installation in a freestanding greenhouse, it is recommended that the floor be set several inches above the finished outside grade of the greenhouse. It also is recommended that when preparing the concrete foundation, the size should be made one inch longer and wider than the actual greenhouse's outside dimensions. Also, the outside edges of the concrete slab should be thicker to provide greater support and to better withstand damage from frost on a freestanding greenhouse. A drain should be located in the center of the greenhouse's concrete slab so that water will drain either into a gravel pit or into a pipe that flows into a drainage area beyond the perimeter of the greenhouse. The gravel pit helps to filter toxins out of the water that may have accumulated on the concrete.

It also is recommended that at least 4 inches of compacted gravel or stone be placed over the top of the subsoil to provide drainage. In addition, placing a 0.24 inches (6 millimeters) polyethylene moisture barrier on top of the gravel or stone will help to keep the concrete slab dry.

1. Build a form out of lumber around the perimeter. The top of the form should be at the height of the finished floor. Adding some reinforcing wire or fiber will help to increase the strength of the concrete slab. Typically, it takes about 24 hours for concrete to set.

2. When the concrete slab has set, remove the lumber forms from around the slab's perimeter.

3. Install insulation board that is 1 to 1 1/2 inches thick vertically around the outside of the foundation. The insulation board should be installed to a depth of 2 feet. This will help to insulate the floor and will also help to keep it warmer in the winter.

4. It also is recommended that whether you are building a concrete slab or wall, you should fasten a two-by-four sill plate. The sill plate is the bottom horizontal boards of a wall where the vertical boards are attached. The sill plate will be anchored to the foundation wall, with the bottom of the sill plate 6 inches above the finished grade.

In the winter, the sill will act as an insulated buffer between the concrete and the greenhouse frame, which will help reduce heat loss just as it does in a residential building. The most common materials used for sills include naturally resistant woods, such as cedar and redwood, and plastic composite timber. The greenhouse base can be attached using concrete anchor bolts that are set within 1-foot of each corner. Additional concrete anchor bolts should be spaced about 4 feet apart. You can purchase these bolts at most hardware stores and home supply outlets. Prices vary but are in the $12 range for a box of 100.

When you have finished securing your concrete base, you can start to consider the advantages of installing a concrete, or permanent, floor in your greenhouse. Permanent floors provide a stable surface that will easily support benches, storage containers, and sinks. Once the floor is installed, there is nothing more to be done.

Of course, the simplest way to install a floor in a greenhouse is to leave the ground bare, and prepare the soil so sections of it can be used for planting. If you leave your ground bare, you will want to take precautions against having a damp or muddy floor in your greenhouse. You do not want diseases and insects to invade the greenhouse and infect and harm the plants growing there. If you plan to have a dirt floor in the greenhouse, it is recommended that you cover it with pavers, flagstones, or raised slats and create a path. That way, it will be more convenient to use a wheelbarrow to move materials inside the greenhouse. Also, because weeds can be a challenge, first lay down a use weed barrier cloth to keep them from growing into the greenhouse.

Gravel is another popular choice for a greenhouse floor. Gravel is very easy to install and it is inexpensive. Applying a layer of gravel that is 3 to 4 inches deep will keep mud and weeds at bay. A gravel floor surface lets you water plants, but allows the water to drain out through the gravel. This process also helps increase the humidity level in the greenhouse.

If gravel is your choice, then you will need to prepare the floor before installation, following these three simple steps:

1. Clear the area completely.

2. Cover it with weed barrier cloth.

3. Shovel gravel or bark on the barrier cloth.

There are advantages to this type of flooring surface. For example, it is adaptable to your gardening needs. The floor can be changed with a shovel, to remove rocks or to add soil for new a bed of plants. There is some maintenance required, such as raking and occasionally replacing the materials.

Slope and Drainage for Your Foundation

Do not permit standing water in your greenhouse. It can breed disease, as well make the greenhouse cooler in winter. During the summer months, standing water can also breed mosquitoes. Wet patches could cause you to slip and fall.

Your solution to these problems is in the slope of the greenhouse floor. The slope should be planned so that water will drain out and away from the building. If the greenhouse has a hard surface such as a concrete slab, the slope should have a 1/8 to 1/4-inch per linear foot drop.

Building with a Greenhouse Kit

If you are uncomfortable working on do-it-yourself construction projects, then a greenhouse kit could be the right choice for you. Most kits are easy to build and come with detailed instructions that will guide you through each step of the process. A greenhouse kit can be purchased online or at a garden center. If the greenhouse you intend to build comes from a kit, carefully read the manufacturer's instructions. Never cut anything unless the instructions indicate that you should.

Before you begin assembling the greenhouse, lay out the parts on a flat surface according to their location in the completed structure. Check the part numbers as you lay them out. All pieces should be identified and properly organized before you fasten them together. If there is a mistake, it will be more difficult to rectify later.

Typically, a kit greenhouse will require a special polyfoam sill moisture-barrier/insulation to be laid on the foundation. Then, the sill or base-plate cover must be installed over the polyfoam and bolted to the concrete foundation. The walls would then be assembled and stood up on the sill or base plate, all of these steps following the instructions that came with the kit for proper, safe, and easy installation.

When you build a greenhouse from a kit, it may be necessary to set up the support posts. The support posts are fastened to the foundation plate and it is recommended that you determine the location for these posts before the foundation is poured. That way, the bolts and tubing for the posts can be inserted into the concrete while it is wet. When the concrete dries, the bolts and tubing will be set. Position the posts in the proper locations where the supports will be installed.

Some greenhouse kits may require that you assemble the aluminum struts and braces for the frame and bolt together the pieces of the pre-drilled frame. Other kits provide pre-assembled walls, roofs, and end walls, in which case all that needs to be done is bolting the top and sides together. Some greenhouse kits only require the glazing to be installed.

Pre-hung doors are a feature of many greenhouse kits. Other kits require you to hang the doors yourself. One option is to use sliding doors set on tracks. They are a good choice because they are durable.

It is recommended that as you assemble the greenhouse kit's different parts, you remember to check that the structure is square. Also, it is a good idea to install any necessary bracing as you work. Spread out the construction over a period of two days. Tighten the bolts on the first day. Then, on the morning of the second day, you will typically find that the bolts can each be turned another half of an inch.

Building Frames

For the purposes of building a greenhouse, or most other buildings, framing provides a stable skeleton on which construction occurs. Studs, which are vertical posts that support a wall, provide a stable frame on which interior and exterior wall coverings are attached.

Frames can be made of wood or metal, and PVC, which is a durable synthetic resin made of polymerizing vinyl chloride, sometimes also used to make flooring.

Many greenhouse kits provide instructions for framing. Typically, you just follow directions to bolt the walls and roof to the sill in the proper order. Most kits include pre-cut components, so the spaces for vents and doors are already made for you.

A do-it-yourself greenhouse construction is obviously more challenging, but it does offer an important advantage: flexibility to customize the design according to the builder's preferences. For a do-it-yourself greenhouse, sections of frame are assembled separately. Then they are attached to a sill, wall, or ledger board, which is a two-by-six piece of lumber attached to the studs with lag screws used in an attached greenhouse. The studs run along an existing wall and, like a sill, provide a surface to nail, typically for the roof rafters. Then plates, which are pieces of two-by-four horizontal supports for attaching the sidewall studs, are attached at the top and bottom.

The lower plate is secured to the sill. Rafters, which form the roof, are laid out across the plate at the top of the wall. Both plates and ledgers run the length of the house. Rafters typically run the width of the building. In a freestanding greenhouse, the rafters are also attached to a ridge board to form the roof's peak.

To help keep the frame square as it is nailed together, it is recommended that you build each section of the frame with the pieces laid out on a level surface. A garage floor or sheets of plywood will work. If necessary, ask for help from family or friends to raise the frames. It will help you avoid dropping and damaging them, or worse, injuring yourself.

Building a Conventional Greenhouse

If you are a competent woodworker, and a do-it-yourself type of person, you may feel comfortable enough to build a conventional greenhouse. There are a number of benefits to constructing a conventional greenhouse yourself. To begin with, there are more features that can be customized according to your own taste and needs. A do-it-yourself greenhouse allows the builder to modify the plans while the building is being constructed.

When building a conventional greenhouse, however, it is important to determine which type of glazing will be used before the work begins. It is the glazing that will determine the type of framing the greenhouse will require. For example, if glass will be used, it can be placed in the greenhouse frame and securely fastened. However, a material such as polycarbonate can be attached over the framework.

A tool you will need to use is a **framing square**. One side of this tool is the same width as a two-by-four. This tool provides a useful measure of width as well as a square. The framing square is used to ensure that the studs are at perfect right angles from the base board.

Obviously, a do-it-yourself greenhouse will require more tools and equipment, because it will take more time to build than if you use a greenhouse kit. Constructing a greenhouse from wood, such as cedar or redwood, may require power woodworking tools, such as a planer or table saw to properly cut the lumber. Although many greenhouse owners who are handy carpenters have these tools, you can also rent them or pay a commercial woodworking company a fee to fashion them in their woodworking shops.

Lumber comes in standard lengths of 8, 10, 12, 14, and 16 feet. When you are estimating how much lumber you will need, add in some extra. The additional wood can be used to fashion braces and attach the coverings. When you plan to construct a wooden building, it helps if you are familiar with standard lumber sizes. These standard sizes will need to be incorporated into the design and construction of the greenhouse. Suppose a design requires a 4-foot, 9-inch wooden structural wall, then the studs would be cut to 4 feet, 4-1/4 inches and would use a sill plate and double cap to get 4 feet, 9 inches. Remember, a stud is a vertical post that is one of the supports of a wooden wall. The problem is that this wastes 3 feet, 7-1/2 inches on each stud. If the walls are modified to measure 4 feet, 4 1/2 inches, then the stud could be cut in half and would use a sill plate and double cap plate to measure up to 4 feet, 4-1/2 inches.

For walls that are lower than 4 feet, 4 1/2 inches in height, you can get two pieces of support wood from each stud. If you size the structure yourself for standard framing lumber, you can minimize the amount of waste and reduce the cost of construction.

Another thing to keep in mind when building with wood — use treated lumber for your greenhouse. It better resists water damage and insect infestation. To increase the wood's toughness, and boost its light reflectivity, paint wood parts with white latex paint. Also, you always want to be sure that you keep the greenhouse frame steady. You can do this by standing on one stud as you nail in the plate to help keep the frame from moving under your feet.

You are now familiar with how to choose a level surface and prepare it for your greenhouse construction. You also learned how to install both a concrete slab for a floor, as well as a gravel surface. These construction basics have set the stage for you to move ahead and begin building the other components of your greenhouse such as the sidewalls, end walls, and roof covering.

Summary

When you are building a greenhouse, the first thing you always build is the foundation. If you do not build a foundation, unless you are making a cold-frame/hoop greenhouse, you are going to have a very unstable structure. The foundation is where the walls secure, and the foundation helps provide you with a solid surface that you can work with. The type of foundation depends on you and what you can afford and build. If you want something simple, then a gravel foundation will work. If you want to be up off the ground to prevent flooding when it rains, then a wooden foundation will be your cheapest option. However, if you want to go the extra mile, the concrete may be the best foundation for your greenhouse.

Chapter 5

Constructing the Lean-To/ Attached Greenhouse

"I don't divide architecture, landscape, and gardening; to me they are one."

— Luis Barragan, Mexican architect

Now that you have begun to lay the foundation of your greenhouse, the next course of action is to begin work on the greenhouse. There are plenty of different styles that you can make, and it really depends on the amount of space you have in your yard. You can either go big or small with your greenhouse, and it all depends on what you want and the space you have. There are several types of attached greenhouses that you can design for your yard. Here we will be going through the simplest form, which is the traditional lean-to that has a zipper door in it. You have built your foundations, whether they are wood, gravel, or concrete. At this point, we will begin to start attaching walls to the sill plates so that we can make your greenhouse. For the next few chapters, we will go over the various types of greenhouses that you can build on top of your foundation.

Why a Lean-To Greenhouse?

If you decide to build a lean-to greenhouse, you are making a good decision in many ways. By building this easily constructed greenhouse, you will be increasing the property value of your home. When there is a lean-to greenhouse attached to your home, the greenhouse actually becomes a usable living space on your property, and you can even add the lean-to to the total square footage of your home. This means that you attach the greenhouse to the side of your house. If you have a door from your house going into the greenhouse, it can count as a room for your home, thereby increasing the property value. In addition, lean-to greenhouses added to a home can improve the structural beauty of your home and even the design.

If you live on a small property and do not have much room for a large garden or large greenhouse, then a lean-to greenhouse is the perfect solution. A lean-to greenhouse can be set up on a property that has limited space.

It is also much easier to install a lean-to greenhouse because of where it is. When you set up a greenhouse elsewhere on your property, the ground may not be level. This means you have to alter the soil so that the ground is completely level for you, which makes it much easier for you to set up your walls and floor. With a lean-to greenhouse, you are using the ground around the house, in contrast to other greenhouses that separate away from the house. One of the walls of the greenhouse will literally be the wall of your home. Typically, this ground will already be level with the home and therefore you will not have to do much in order to alter the soil or the level of the ground, which makes things much easier for you.

CASE STUDY: GREENHOUSE ENTHUSIAST

Carol Van Klompenburg
The Write Place, Owner
www.thewriteplace.biz

Carol Van Klompenburg, owner of The Write Place, has been gardening on the side for years. "Our first greenhouse was a sun porch on the east side of our house. My husband and I built it because we wanted more indoor sunshine," she said. "We had no south-facing windows on our house." As she began to start flowers under grow lights in the basement, she realized that the sun porch would be a better location — combining grow lights and sunshine. So, she began placing tables along the south and east windows, and starting flowers. However, soon Van Klompenburg realized that for starting seeds, she needed to supplement with grow lights.

After her first successful attempt at a greenhouse, the second greenhouse she purchased was a translucent pop-up plastic greenhouse. She said she purchased this for two reasons: "I wanted more space to start flowers and a friend offered to sell it to me at a good price. She had bought it several years earlier and had never used it," said Van Klompenburg.

Van Klompenburg liked the functionality of her plastic greenhouse; however, she and her husband — who is a recently retired mechanical engineer with the necessary construction skills — did not like the appearance of the plastic tent, and they knew it would not last forever. "We had seven sliding glass doors and a front door we had removed from our house," she said. "We both like to reduce, reuse, and recycle and wanted to make use of the materials we had removed from our home and replaced with more energy efficient ones." And although this second greenhouse attempt was not what she thought it would be, she said that it was her favorite one so far. "The pop-up tent is my favorite at the moment," she said. "It is a separate space dedicated to plants and does not take over my living space in the spring."

Not only did Van Klompenburg make her home more energy efficient, but she loved the way the greenhouses helped extend the growing season, both in the spring and the fall. "By February, I am eager to start working with plants and starting indoors in the greenhouses makes that possible," she said. She also loves her greenhouses because they enable her to have a May plant sale for other flower gardeners — and this sale is what funds her flower purchases for the new season. Although her greenhouse is not essential for her flower gardening, it is definitely beneficial for her and it is essential for starting plants early and for enabling her to sell plants and raise gardening funds.

As Van Klompenburg's greenhouse expertise expanded, and she discovered a pleasure in watching flowers grow, she made a few different discoveries along the way. "I discovered that I needed to supplement the light in the sun porch for starting seeds. Otherwise the seedlings were too spindly, not sturdy enough," she said. She also discovered, when she owned her second greenhouse, that the pop-up tent needed to be well-tethered and secured. She said the first time she put it up, two sides collapsed due to strong wind. Another thing she discovered was that a high humidity greenhouse area was more prone to pests and disease, and it was important that she needed to be more vigilant than in her outdoor beds.

Things to Consider for Your Project

When you are building a lean-to greenhouse, you have made a good choice especially if this is your first greenhouse project. The simplest type of greenhouse is the lean-to greenhouse and you can easily fit it into your backyard. Also, they are incredible easy for you to assemble. Of course, they are not foolproof to build, so there are some things that you will need to consider when you are putting together a lean-to greenhouse.

First, most lean-to greenhouses come with a back panel and this attaches to a wall of a house or an outbuilding. When you are building a lean-to greenhouse, it is very important that you build the greenhouse with a wall to rest against the wall of the house. There are two big reasons for this; heat loss and dampness. The heat from a home next to the greenhouse mixed with the heat from the wall of the greenhouse can actually cause dampness on the wall of the greenhouse. The humidity may therefore be higher than what you were expecting, which can cause problems for you and your plants.

Next, when you are putting together lean-to greenhouses, especially from a kit, make sure to read all the instructions and precautions. These instructions and precautions will typically cover the tools you need, how to safely use the tools, and how to put together the greenhouse. You also need to consider what type of material you are going to use for your lean-to greenhouse. You can typically choose between wood, vinyl, brick, and stone, with most people choosing wood, because it is the easiest to work with. Considering some protective coating is also very important. If you are using brick or stonework as the walls for your lean-to greenhouse, then having plastic over this will help keep mold from eating through the brick and stone when it is hot. In this case, it is highly recommended that you use something like silicone rubber, which you can get from a hardware store and you can use it to seal edges.

You need to think about what your surroundings are going to be like. If you live where there is plenty of pollution or rain, then you need to have a good wood construction for your greenhouse, as well as glass covering it. However, you can use fiberglass, which is much cheaper. Thankfully, wood is very cheap and it will not cost much to buy enough to build your entire greenhouse and then some.

Now, if you live somewhere that has plenty of sunny weather, you can use something like polyethylene, which is very cheap. If you can spend more than others can, then you may want to consider aluminum for your greenhouse. Because you are essentially building your greenhouse out of metal, it is going to cost several hundreds of dollars more, and that is something you should consider if you are going this route.

Ventilation is also very important. The greenhouse needs airflow going into it and with a lean-to greenhouse a window or a flap will be enough to provide you with enough ventilation.

Irrigation is not something you have to worry about with your lean-to greenhouse, because you probably are going to have the lean-to greenhouse attached to your house. This means you will have close access to water. In regards to heating, this is something to consider, especially if you are at northern latitudes. It is important to have a material that will help keep heat in your greenhouse because using a greenhouse can help extend your growing season by several weeks. Plastic sheeting that you can buy off the roll at hardware stores will be inexpensive, costing 50 cents to a dollar per foot typically.

Do not forget to make the lean-to greenhouse look good. It attaches to your home and, therefore, you should remember to make it look good. If you want to increase the value of your home with this greenhouse, find ways to make it look like an extension of the house, rather than something that serves as an eyesore.

The most common size for a lean-to greenhouse is typically 8' x 12'. In addition, ensure that your roofing is not higher, or wider, than the roofing of your home. Your lean-to greenhouse needs to have windows, but you have to keep in mind where you are putting these windows. Putting a window on the top of the lean-to greenhouse is a good idea, but you do not want it to be too large. There will be weight on the roof when it snows or rains and you do not want the window to collapse because of that weight on the window.

There are some disadvantages to using a lean-to greenhouse. First, if you love to garden and you want to use your garden as much as possible, then a lean-to greenhouse is not always the best option because it is rather small. Lean-to greenhouses are smaller, because they are at a slant against the wall of your house or another building. This creates limited space because you can only have a wall that is so long, otherwise the angle will be too steep to allow you to work in the greenhouse.

With your lean-to greenhouse, there are some problems with the fact that your plants may not get all the sunlight they normally would in a garden, because the lean-to greenhouse is on one side of the house, causing the house to shade the greenhouse at times. This can cause the plants to discolor and they may not grow as large as they would outside.

The following sections of the book will explore different plans for your lean-to greenhouse. Choose the one that appeals to you the most and use those instructions to build your lean-to greenhouse.

Lean-To Greenhouse Plan No. 1

This first plan is designed for the gardener looking for a one of a kind styled greenhouse. As you will notice — once you begin reading the instructions — there are no actual measurements to follow as you begin to mark-off the area for your building. You will map out your own perimeter so as to create a lean-to greenhouse customized for you. This plan will give you more freedom when designing your structure.

1. Measure out the structure on the ground for the size that you want.

2. Trench out the perimeter you have created. This is where you are going to be putting down your foundation. You can put down a foundation of cement, wood, or just dirt. It is up to you how much money you may have for this endeavor.

3. Frame up your walls using two-by-twos. The best wood to use for this is typically spruce and cedar, but use what you can afford and ask at the hardware store what wood is best for your climate.

4. Screw in a two-by-two board that is as long as the width of the greenhouse against the wall. This serves as your vapor barrier discussed earlier.

5. Frame the front wall of your lean-to greenhouse. Remember, your front wall must be shorter than your back wall. If your back wall is about 8 feet high, then the front wall should only be 6 feet high. This will then allow water and snow to slide off the roof, thereby keeping moisture from pooling above your greenhouse and saving it from eventually collapsing.

6. Begin doing the vertical framing of your front wall by having the studs about 24 inches from each other. At the front wall, ensure that you leave a 30-inch space for the door.

7. At this point, you can begin framing the sides to the right height. Remember, you will have to cut each stud at an angle at its top. The reason for this is that the horizontal board on the top will be running at an angle from the higher back wall to the shorter front wall. To ensure things fit properly, you will need to cut the sidewalls to the right height and the right angle.

8. Once the wall studs are in, you can install the horizontal supports. These will run across the roof from side-to-side to provide extra strength. Have one stud run from the back wall to the front wall, in the middle of the greenhouse. This will give additional support to the other horizontal supports.

9. At this point, you will need to put up a wire mesh that measures half-an-inch by half-an-inch. Cut it to fit the roof and staple it into place. You are putting this on there to provide support for the polyethylene plastic that you are putting on your roof. Without the mesh resting on the support studs, you will find the plastic begins to sag and rip. It is very important to be as accurate as possible with

your construction. If you start to cut too long or too short, it could result in your greenhouse looking shabby.

10. Staple the plastic down all three sides of the lean-to greenhouse and to the roof of the greenhouse. Do not staple against the wall of the house where the greenhouse rests. To create an air seal, you will need to overlap seams and staple these seams down.

11. Purchase a greenhouse zipper door from the hardware store and install it in the 30-inch space you created at the front of your greenhouse.

Lean-To Greenhouse Plan No. 2

Unlike plan No. 1, this plan will give you the exact measurements you need if you are someone who is not as confident in your ability to mark-off the area for your greenhouse. You will also notice that this plan is specifically for wood flooring inside your lean-to greenhouse

1. Mark out a 5 x 8 floor plan for the lean-to greenhouse.

2. Lay down a wood foundation by digging the foundation trench. Your sidewalls will be 5 feet long, and the back and front walls will be 8 feet long.

3. Lay down logs that measure 8 x 8. If you go parallel from the sidewalls, you will need eight-by-eights that are 5 feet long and you will need 12 of them. If you decide to go parallel to the back and front walls, then you will need eight-by-eights that are 8 feet long and you will need 7-1/2 in total. Naturally, it is much easier to go parallel, because you do not have to cut lengthwise.

4. Once you have laid down the wood foundation, you can begin framing. You will want to use wood that is either 2 x 2 or 2 x 4. What you need to do is to use a level and draw a line from the soil

to 7 feet up on the back wall. Put in a board along that line. Now, using a level, have another line drawn across the back wall. Put the board in along that line, as well as at the other side of the board.

5. Build the outside wall first. Cut a top and bottom plate for the wall that is 8 feet long. The boards will be marked and installed every 16 inches. Cut each stud to the proper 6 feet high.

6. Once you have built the outside wall, attach it to the foundation boards. Keep the wall up by propping up some boards against it. Using a level, check to ensure that the wall is straight. At this point, nail a rafter on each end of the wall to the board on the back wall. Then, put in the studs at the top every 16 inches.

7. Now that you have created the back and front walls, you can put in the sidewalls.

8. On one of the walls, probably the west, you will have an opening that will be 20 inches wide. This is where you are going to put the door. Nail the rest of the studs 16 inches apart.

9. The doorframe itself should be 19-1/2 inches wide and 5 feet, 11 inches high.

10. Nail in the studs for the other wall so that you can install both walls and attach them to the back and front walls.

11. Now, you will install vent doors. These vents will be 14-1/2 x 12 inches. You will put two on the same wall as you have the door and one on the roof. Make sure that you frame in these vents about 1-foot off the ground. The roof vent is center. You can put in the vents using two hinges each and some vents you buy at the hardware store.

12. At this point, cover the greenhouse with plastic or fiberglass panels. You can use a staple gun to put the fiberglass down, and the panels can be installed using just nails.

Greenhouse Plan
Lean -To #2
5 FT x 8 FT

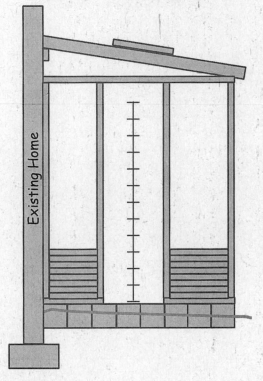

Existing Home

Bill of Material

Item	Qty	Description	Item	Qty	Description
1	8	8x8 timber x 8'-0"	9	2	2x4 x 5'-4"
2	6	2x4 x 5'-9"	10	2	2x4 x 1'-5"
3	1	2x4 x 8'-0"	11	2	2x4 x 1'-8"
4	2	2x4 x 7'-5"	12	1	roof vent
5	13	2x4 x 6'-0"	13	2	zippers
6	1 box	3" long 10d nails	14	2	wall vents
7	2	2x4 x 1'-9"	15	1 roll	UV resistant polyethelyne
8	2	2x4 x 2'-0"			

Step 1: Assemble Timber Foundation:

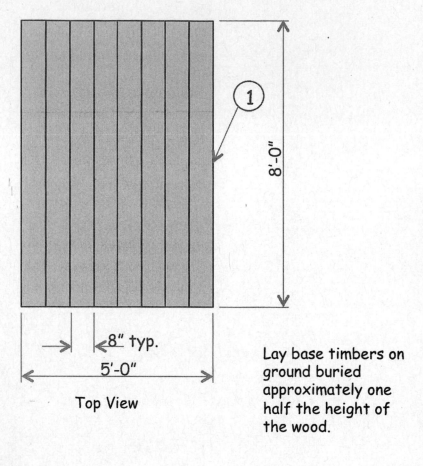

①

8'-0"

8" typ.

5'-0"

Top View

Lay base timbers on ground buried approximately one half the height of the wood.

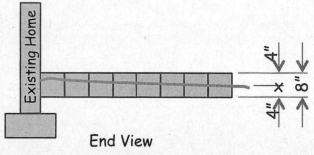

Existing Home

4"

8"

x

4"

End View

Step 2: Assemble Frame:

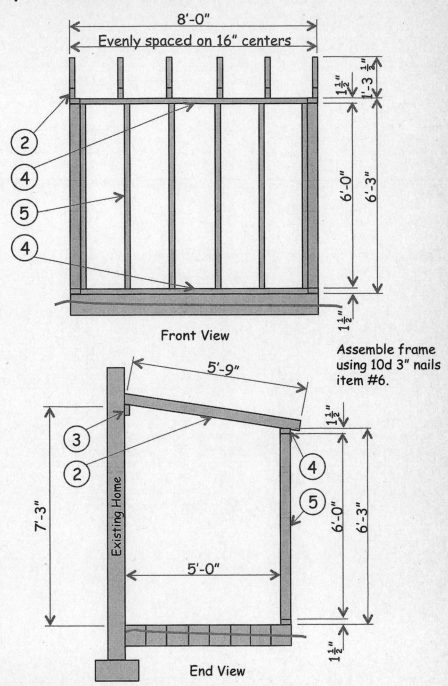

8'-0"

Evenly spaced on 16" centers

2
4
5
4

$1\frac{1}{2}$"
1'-3$\frac{1}{2}$"
6'-0"
6'-3"
$1\frac{1}{2}$"

Front View

5'-9"

3
2

Existing Home

7'-3"

5'-0"

4
5

$1\frac{1}{2}$"
6'-0"
6'-3"
$1\frac{1}{2}$"

End View

Assemble frame
using 10d 3" nails
item #6.

Step 3: Assemble End Frame:

Assemble end frame
using 10d 3" nails
item #6.

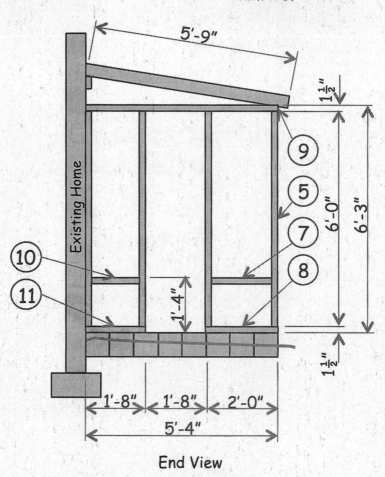

End View

Step 4: Final Assembly:

Install vents and cover structure
with UV resistant polyethylene or
fiberglass panels.

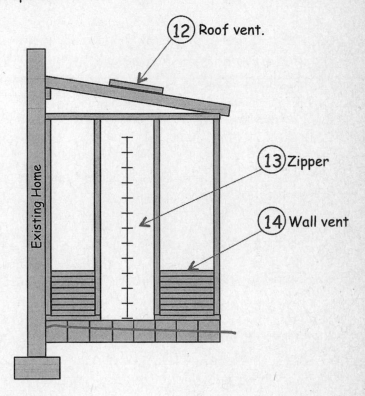

End View

Lean-To Greenhouse Plan No. 3

This lean-to greenhouse is a few feet larger than the previous plans, but it still has the same relative design, in that the roof and front wall create a 90-degree angle. You will also notice that, unlike with the plan No. 2, you can use any type of material for the foundation.

1. Mark out the perimeter of your lean-to greenhouse. This lean-to will be 8 x 12 feet. So, there will be 8-foot sidewalls and 8-foot back and front walls.

2. Use the large eight-by-eight timbers, dug to half their depth, to create a foundation that helps the greenhouse rest off the ground.

3. The back of the greenhouse will be 8 feet high and the front wall will be 6 feet high. You will need to install the ledger board, which is the board that serves as the top of the back wall. This will be 8 feet high and run horizontally 12 feet across the back wall.

4. Install the sideboards that will serve as the sides of the back wall and attach this to your sidewalls.

5. Frame out the front wall so that it is 12 feet across and perfectly parallel with your back wall. You should have your studs placed every 2 feet from one end to the other of the front wall. Make sure you keep space for vents. You will be installing vents between the four middle studs in the middle of the wall.

6. Once you have framed out the front wall of the greenhouse, you will need to attach it to the ledger board of the back wall. Have someone hold up the front wall while you run an 8-foot 3-inch board that runs down at an angle from the back wall to the front wall. Do this on both sides of the front wall. You should now install your front wall.

7. At this point, you will frame the top of the greenhouse. Frame with the two-by-fours every 2 feet along the top of the wall. Make sure that your roof rafters rest on top of the top board of your front wall. This will help ensure there is proper run off from the roof. Leave room for the four vents in between the four center studs at the back of the roof that are against the ledger board.

8. Once you have installed the boards that run from the back wall to the front wall, you will need to install the roof boards that run from the top side board on the one end to the other. These run parallel to the front and back walls, and you need to cut 2 feet each. You will place one 4 feet from the back wall and one 4 feet from the front wall. Therefore, you will only have two of these blocking studs in each section. Keep them all in line and run them from the one sidewall to the other.

9. At this point, you can begin working on the sidewalls. You will build these slightly differently than in the other plans. For the side with the door, you will put one stud 2 feet from the back wall and one 2 feet from the front wall. These two studs will only be 6 feet high. At the top of them, you will run a board that is 8 feet long from the back wall to the front wall.

10. At this point, you need to make sure you make room for the door. On the side of each stud (the sides that face the other stud), you will put another stud 8 inches in. In between those two studs that are now 8 inches apart, you put a small two-by-four horizontally in the middle. Do the same with the other stud. Now, in between these four studs, you will have your door space.

11. At this point, you will frame in the other wall. This one is easier to frame. Put 6-foot high studs in every 2 feet on this wall and attach them to both the top and bottom boards. Above the top board, install the two studs the same way you installed them on the other wall, 4 feet from the front and back walls and 4 feet from each other.

12. You need to install the covering now. You can just use plastic, but if you want it to look better, you can install rigid plastic or fiberglass. The plastic sheets should be directly next to each other but they should not overlap. As well, make sure that you drill holes into the wood where you will be hammering in the nails to secure the plastic.

13. On the roof, install the fiberglass sheets on the rafters using the same method that you used on the sides. However, on the sheets on the roof, you can overlap the sheets because you want to have a complete seal that will not allow water into the greenhouse. Also, make sure that the sheets hang 2 inches over the top of the edge of the roof.

14. Now all you have to do is install the door that you have purchased into the space where you allocated room for the door.

Greenhouse Plan
Lean -To #3
8 FT x 12 FT

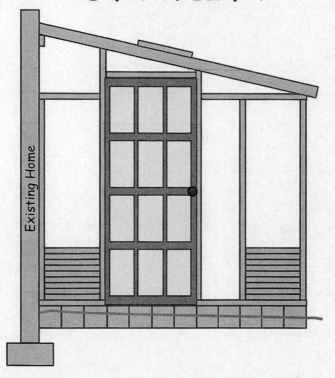

Bill of Material

Item	Qty	Description	Item	Qty	Description
1	12	8x8 timber x 12'-0"	11	2	2x4 x 1'-10 1/2"
2	7	2x4 x 8'-9"	12	2	2x4 x 2'-8"
3	1	2x4 x 12'-0"	13	2	2x4 x 7'-9"
4	2	2x4 x 11'-9"	14	2	2x4 x 7'-0 3/4"
5	7	2x4 x 6'-0"	15	2	2x4 x 2'-0"
6	1 box	3" long 10d nails	16	1	roof vent
7	2	2x4 x 3'-2 1/2"	17	2	wall vents
8	6	2x4 x 6'-0"	18	1	prehung door
9	4	2x4 x 1'-9"	19	1 roll	UV resistant polyethylene
10	2	2x4 x 3'-4"			

Step 1: Assemble Timber Foundation:

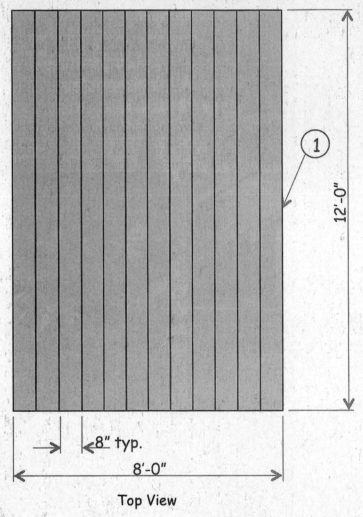

8" typ.

8'-0"

12'-0"

1

Top View

Lay base timbers on ground
buried approximately one
half the height of the wood.

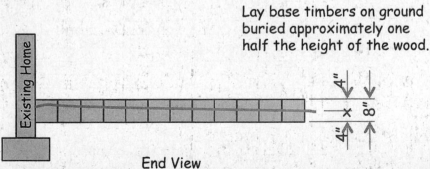

Existing Home

4"

8"

4" x

End View

Step 2: Assemble Frame:

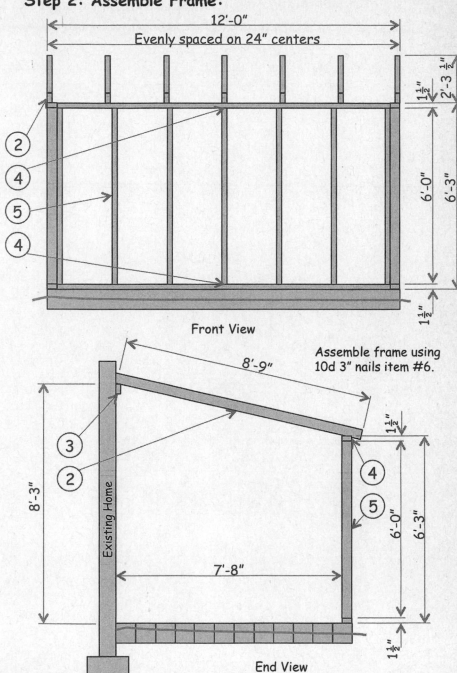

12'-0"

Evenly spaced on 24" centers

1 1/2"

2'-3"

2 1/2"

2

4

5

4

6'-0"

6'-3"

1 1/2"

Front View

8'-9"

Assemble frame using
10d 3" nails item #6.

3

2

Existing Home

1 1/2"

4

5

8'-3"

7'-8"

6'-0"

6'-3"

1 1/2"

End View

Step 4: Final Assembly:

Install vents and cover structure
with UV resistant polyethylene or
fiberglass panels.

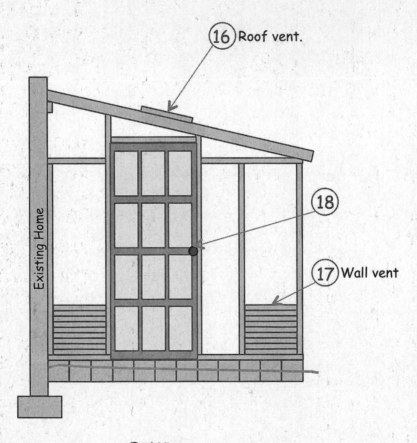

End View

Lean-To Greenhouse Plan No. 4

This greenhouse is not like the others, which have a slanted roof and front wall that is at a 90-degree angle to the ground. Instead, this greenhouse has both a slanted roof and a slanted front wall, but is still a lean-to. This allows for a greater amount of light to come into the greenhouse, which is important in the winter.

1. Draw out a plan for an 8 x 12 greenhouse that has a foundation of wood or concrete, depending on your preference.

2. Install the bottom sills (boards) of the greenhouse by securing them into the concrete or into the large timber logs that serve as the bottom of your greenhouse.

3. Install the ledger board 8 feet above the ground against the house. This board should be 12 feet across. After you install this, install the sideboards that are 8 feet high and connected to the ledger board.

4. At this point, you will need to put together the front wall. The front wall will be 12 x 7 feet. Therefore, it will be 7 feet tall and 12 feet wide. Build the wall separately and have seven studs that are 2 feet apart. Each stud needs to attach to the bottom of the frame at a 60-degree angle so you get the proper angle to connect to the rest of the greenhouse. Make room for vents in between two of the studs. The vent will need one two-by-four about 4 feet above the bottom board to help create a brace.

5. Brace the wall up so that it rests at 60 degrees to get the most sunlight possible. At this point, attach a 6-foot, 1-3/4-inch board to the ledger board on each end, and then attach the front wall to this board.

6. The sidewalls will be relatively easy to build now. Run a board horizontally 6 feet above the ground from the back wall to the front wall. Then, attach studs in that leave room for the door on

one side. Remember, you will have to cut room for the door on the wall. Also, leave room in the top corner of the sidewall for a vent.

7. At this point, you need to put in the rafters. Attach rafters to the front wall by the use of a bird's mouth joint. Install rafters every 2 feet, just the same as the front wall's studs. In fact, done right, everything should join together at the bird's mouth joint where the rafters meet the top of the front wall and the studs of the front wall.

8. Next, once you have installed your door, you need to install the covering for the greenhouse. Cover the roof with shingled plywood. Leave room for vents on the top. On the front wall, install large windows, or secure clear plastic in place. On the sides, install acrylic panels or fiberglass, which allows sunlight to still come through.

Mini Lean-To Greenhouses You Can Buy

If you just do not want to build a greenhouse, or you do not have room for a large one outside, you can try these different types of mini lean-to greenhouses instead. These types of greenhouses are not pre-built; they require assembly, so be sure you allot a day or two to complete your project.

Four-tier mini greenhouse.
Photo courtesy of Juliana America, LLC.

- **A four-tier mini greenhouse** allows you to have your greenhouse nearly anywhere. You can put this on your deck because it essentially stands as a shelf, usually about 6 feet high and 2-1/2 feet wide. One of the great benefits of this mini lean-to greenhouse is that if you are growing a big plant, you can remove a shelf in order to give yourself some extra room. Another benefit is that it is small enough that you can move it indoors when the weather gets cold so you can keep growing food during the winter. Essentially, it will look like a shelving unit in a room that gets plenty of sunlight.

- **A European style greenhouse** is a small greenhouse that attaches to the side of your house. It includes a vent on the top and is typically about as wide as the coolers you will see inside of a convenience store. Usually these lean-tos will be about 4 feet wide in total and will leave you ample room to grow several plants of varying heights. The only problem here is that it is not small enough to move easily into your home when the weather gets cold.

- **A carport lean-to greenhouse** is a small lean-to that you can install easily from a kit. It measures about 10 feet long, 4 feet wide, and 7 feet high. Complete with polycarbonate panels and sliding doors, its aluminum frame makes it nice and light, but it is still too large to move into your home when it gets cold during the winter.

Carport lean-to greenhouse.
Photo courtesy of Juliana America, LLC.

Simple Mini-Greenhouse

Here is an example of a mini greenhouse with fresh garlic growing in front. The gardeners might chose this mini greenhouse kit if they only need enough room to grow fruits and vegetables for two.

The simple mini-greenhouse is perfect if you do not have much room but you want to grow some vegetables that can grow in pots. Some of the best vegetables to grow in a mini-greenhouse include tomatoes, peas, beans, carrots, and onions. If you have a small backyard, the mini-greenhouse may be the perfect type of greenhouse for you.

A-Frame Greenhouse Plan No. 1

This plan will allow you to build a simple, modified A-frame greenhouse out of plastic PVC pipe. As Chapter 3 mentioned, PVC pipe is used for the construction of more small scale greenhouses. For this project, you will need the following materials.

- (16) 3-foot pieces of PVC pipe

- (8) 6-foot pieces of PVC pipe

- 8 straight PVC pipe fittings with three openings

- 8 corner PVC pipe fittings with three openings

- PVC cement

- Adhesive

- Plastic sheeting

1. Attach two 3-foot long PVC pipes together with one straight pipefitting. Secure the pipes together within the fitting using PVC cement.

2. Continue to repeat this process until you have created eight straight pipes that measure 6 feet long. These pipes will serve as the base and the top for your greenhouse frame.

3. Attach four of the 6-foot long pieces with the three opening corner fittings. Insert pipes into these openings and leave the third opening facing upwards.

4. Insert one 6-foot length of PVC pipe into the third opening on each corner and each straight fitting into the base of the greenhouse.

5. Put in the 6-foot lengths of pipe on top of the supports you have made using the rest of the corner fittings.

6. Cover up the portable greenhouse with plastic sheeting and use the adhesive to join it to the PVC pipe.

7. Cut an opening for the door for your greenhouse and glue the plastic base from one side of the slit to the corner.

Greenhouse Plan
Mini A-Frame PVC
6FT x 6FT

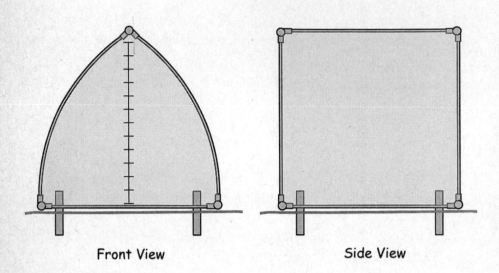

Front View   Side View

Bill of Material

Item	Qty	Description	Item	Qty	Description
1	6	1" PVC corner fittings	6	4	metal pipe straps
2	5	1" PVC pipe x 6'-0"	7	1 box	#8 wood screws x 2" LG
3	1 can	PVC adhesive	8	1 roll	UV resistant polyethylene
4	4	1" PCV pipe x 8'-0"	9	1	6 FT long zipper
5	4	2x2 wood stakes x 18" LG			

Step 1: Assemble Base Frame:

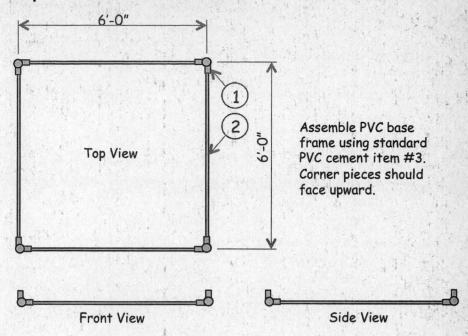

Top View

Front View

Side View

Assemble PVC base frame using standard PVC cement item #3. Corner pieces should face upward.

Step 2: Assemble Top Support:

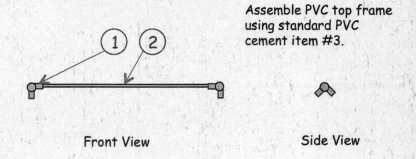

Front View

Side View

Assemble PVC top frame using standard PVC cement item #3.

Step 3: Install Vertical Legs:

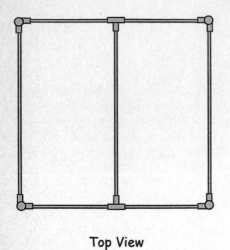

Install vertical legs item #4. Secure with PVC cement item #3.

Top View

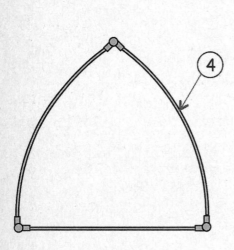

Front View

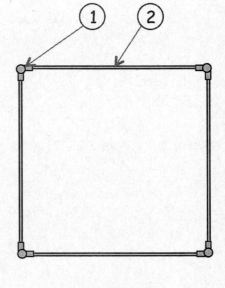

Side View

Step 4: Final Assembly:

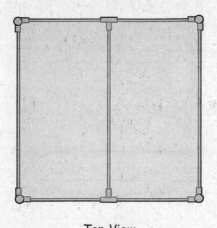

Top View

Anchor frame to ground using 4 stakes item #5, metal straps item #6, and wood screws item #7. Cover frame with UV resistant polyethylene item #8. Install zipper item #9.

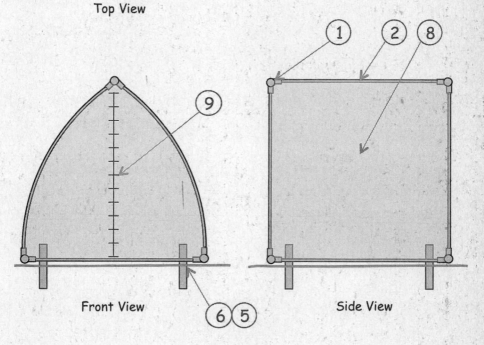

Front View   Side View

Summary

If you are building a greenhouse on your property and you want to go with a simple plan, or you are limited for space, then building a lean-to greenhouse, may be your best options. These greenhouse shapes were in use for centuries and they work quite well. With the lean-to greenhouse, if placed in the right spot, you can have a lot of sunlight getting into the greenhouse. Using your house as a wall gives you a sturdy structure to build the greenhouse onto, and that makes it easier to build.

Chapter 6

Constructing an A-Frame Greenhouse

"Because of technological limits, there is a certain amount of food that we can produce per acre. If we were to have intensive greenhouse agriculture, we could have much higher production."

— Ralph Merkle, Cryptographer

One of the most common types of greenhouses that you will see is an A-frame greenhouse. This type of greenhouse usually looks like a triangle, but there are also forms that look like a house, as discussed in Chapter 1. They can be found as stand-alone structures, as well as attached to a home.

The A-frame is actually very easy to build and looks wonderful year round. They can raise the property value of your home and can be placed anywhere within the yard. It is also very easy to add a door to the greenhouse, because of the high walls and the simple fact that these greenhouses are so simple to make. You have ample head room in the middle of the greenhouse but with the sloping sides of the A-frame greenhouse, you are limited to a certain height for stacking your plants along the walls of the greenhouse. For example, if you have an A-frame greenhouse that is 10 feet wide, then you will have a peak that reaches to 9 feet tall, with the walls at an angle of about 60 degrees. One benefit to this high peak is that you can put hanging baskets down the middle of the greenhouse in order to add to the amount of plants within the greenhouse. You cannot do this with other types of greenhouses. In a typical modified A-frame greenhouse, if it is 10 feet wide, the roof will only be 6 to 7 feet tall. That is not enough room to hang plants above you unless you are shorter than 5 feet.

Further, on the topic of space, you do have the space to put up taller trees and shrubs within an A-frame greenhouse, which is something you may not be able to do in another type of greenhouse. However, under the benches you lose plenty of space, because there is not much height and the angle of the walls cuts down on space that could be used under the counters.

The classic A-frame greenhouse, with its slanted sides, is very strong and can easily get rid of snow, which will simply slide down the walls. If the angle of the slant is above 60 degrees though, there will not be enough structural support and the greenhouse could collapse under the weight of snow. However, it is going to be more expensive to heat when the weather is cold. The modified A-frame greenhouse has a gabled roof and looks like a house. It has more efficient use of space than a classic A-frame greenhouse and does not cost as much to heat, but there can be an accumulation of snow on top of the greenhouse which puts stress on the roof.

Materials to Choose

Choosing the material for your greenhouse is very important, especially with a modified A-frame greenhouse or classic A-frame greenhouse. The reason for this is that you are essentially building a small one room house. You want to make sure that the material you use will be strong enough to support the weight of a roof, as well as glass windows if you use them for paneling. In Chapter 3 you learned the types of material you can use for your greenhouse's structure. The following list is simply a reminder of previously discussed material, along with a few additions to keep in mind when deciding on the frame for your greenhouse.

- **Aluminum** – This is a very popular material used in greenhouse construction because it is inexpensive, does not rust, and is easy to assemble. However, aluminum is not a good insulator for your greenhouse. This means that any heat produced inside your greenhouse will escape.

- **Steel** – This frame material is strong, as it can withstand harsh weather but may not hold up well against the sun unless treated with rust resistant paint to prevent rusting and corrosion. Steel is a good insulator and it supports glass panels well.

- **Plastic** – Unlike aluminum, plastic provides good insulation within your greenhouse. It is also a very strong material. The only disadvantage of using plastic as a frame is that there may be some warping over time; the plastic may change shape due to heat from the sun.

- **Wood** – This material is extremely popular due to its durability and functionality. However, wood has the potential to harbor diseases and create mildew that could affect plant life within your greenhouse.

- **PVC** – This material is also a popular option for individuals wanting to build a hoop-style greenhouse. The downside of using PVC pipe for your greenhouse is that it is not the most stable material, but it does provide insulation and can be long-lasting.

- **Solexx** – This frame is a combination of two other usable greenhouse materials — PVC and steel. It is also a material used to construct bridges, which makes it a very strong frame for your greenhouse. It has been proven to endure severe weather such as wind and snow.

A-Frame Greenhouse Plan No. 1

This is pretty much the simplest type of greenhouse you can construct because it is just four walls with an A-frame roof and the angles are very easy to put together. This greenhouse is perfect for someone who does not have much experience building.

1. Get two-by-twos that are 8 feet in length. You will also need about 128 square feet of chicken wire. Make sure you also get 128 square feet of UV-stabilized polyethylene.

2. Determine where you want to put your greenhouse. You can put the greenhouse directly in your garden if you have the room. One of the great things about a classic A-frame greenhouse is that it does not require a foundation, so you can literally put it anywhere on level ground.

3. Lay down the lumber to create a square, and then use brackets and screws to secure the pieces of lumber together.

4. Put the chicken wire over the lumber and attach the wire to the square you have created. Use a staple gun to secure it and make sure that the chicken wire is not loose on the frame; it should be very tight.

5. Put the required amount of polyethylene over the mesh on the square frame you have created.

6. For the next side of the greenhouse lay down the lumber to create a square, and then use brackets and screws to secure the pieces of lumber together. Put the chicken wire over the lumber and attach the chicken wire to the square you have created. Use a staple gun to secure it and make sure that the chicken wire is not loose on the frame; it should be very tight.

7. Put the tops of the two walls together, forming two sides of a triangle. To ensure you can position the A-frame greenhouse, use door hinges on the part where the two sides join. Use one door hinge 2 feet from one side and a door hinge 2 feet from the other side.

8. Raise the A-frame greenhouse so that it forms into a triangle.

9. At this point, nail in 2-foot long pieces of rebar halfway into the ground. Secure the greenhouse to the rebar using pipe straps.

10. Now, you can plant the seeds directly into the ground underneath the greenhouse. As the weather warms, the interior of this greenhouse will get more heat than other areas of the garden. Use this for plants that require hotter temperatures like watermelons.

Greenhouse Plan
8 FT x 8 FT A Frame
2x3 Wood Construction

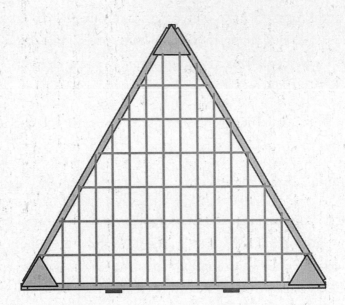

Bill of Material

Item	Qty	Description	Item	Qty	Description
1	4	2x2 x 8'-0"	7	1	2x2 x 8'-0"
2	1 box	#8 screws x 2 1/2" LG	8	4	2" hinges
3	21	1/2" plywood x 8" x 8"	9	1 roll	UV resistent polyethylene
4	1 box	#8 screws x 1 1/2" LG	10	1 roll	chicken wire
5	6	2x2 x 8'-0"	11	4	1/2" dia rebar x 2'-0"
6	10	2x2 x 8'-0"	12	4	latch

Step 1: Assemble Base Frame:

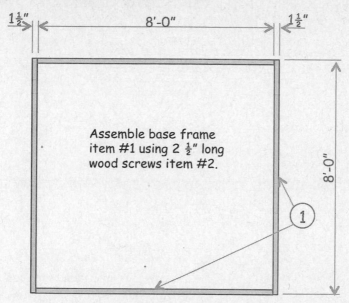

1½"   8'-0"   1½"

Assemble base frame item #1 using 2 ½" long wood screws item #2.

8'-0"

1

Step 2: Assemble Hinged End Frames:

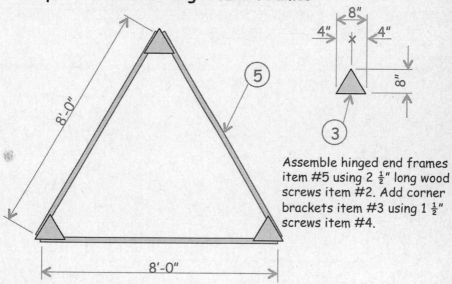

8'-0"

5

8'-0"

8'-0"

8"

4"   4"

8"

3

Assemble hinged end frames item #5 using 2 ½" long wood screws item #2. Add corner brackets item #3 using 1 ½" screws item #4.

Step 3: Assemble Mid Supports:

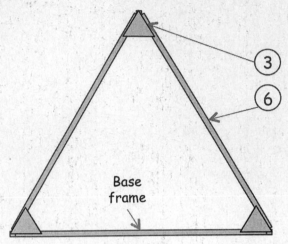

Front View

Attach mid supports item #6 to base frame using 2 ½" long screws item #2.

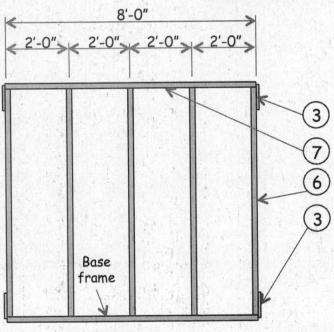

Side View

Step 4: Final Assembly:

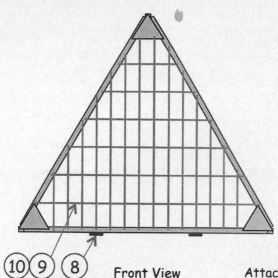

Front View

Attach end frames using 2" hinges item #8. Cover End frames and main structure with chicken wire item #9 and UV resistant polyethylene item #10.

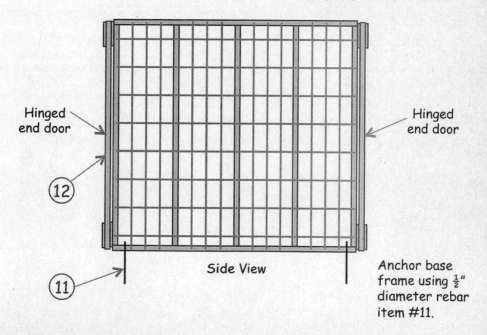

Hinged end door

Hinged end door

Side View

Anchor base frame using $\frac{1}{2}$" diameter rebar item #11.

A-Frame Greenhouse Plan No. 2

This greenhouse plan is a little more intricate, in that you must construct a door and a foundation. This plan is for more experienced builders, but if you would like to try your hand at something more challenging in the way of greenhouse, take a shot. And as with any building project, please take the necessary safety precautions.

1. Choose a site for your greenhouse. Once you have chosen the site for the greenhouse, lay down your timbers to construct what will be the foundation and base of your greenhouse. The timbers you should use are those that are often used as fence posts, because they are sturdy and heavy.

2. Secure the base to the ground by hammering pegs around the base and securing the base to the pegs.

3. Using two-by-twos or two-by-fours, begin to build the sidewalls. On a flat piece of ground, secure 4-1/2 feet high studs an equal distance apart to 10-foot long top and bottom boards. Do this for both sides.

4. Stand up both sidewalls and prop them in place on top of the base you have created. Using galvanized nails, secure the bottom plate (board) of the sidewalls to the baseboard.

5. Now, you need to make the roof frame. This is a bit more difficult because you have to get the peak and angles right, so that the top of the roof is symmetrical and straight in relation to the rest of the greenhouse. In total, you will need to cut five rafters that are 6 feet long, five rafters that are 4 feet long and five uprights that will be 1-foot long. The 6-foot long boards will need to have a 30-degree angle on the left side and a 60-degree angle on the right side. The 4-foot long boards will need a 60-degree angle on the left side and a

30-degree angle on the right side. The 1-foot long boards will need two 60-degree angles cut.

6. Cut out five triangular gussets with plywood that is a quarter of an inch thick so that each side of the triangle gusset measures 1-1/2 feet. Each side angle for the triangle should be exactly 60 degrees.

7. You will need to construct the roof pattern so that the 6-foot rafter is on your left and the 4-foot rafter joins the 6-foot rafter about a quarter of the way down. The upright will then go from the 4-foot rafter up to the end of the 6-foot rafter. This will all be secured together using flathead nails with the gusset. Use the pattern below for the roof.

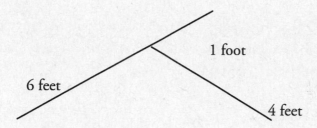

8. Do this five times to create five rafters joined together. At this point, you will lift up the roof frames and secure them to the tops of the sidewalls. Secure these in place to the tops of the wide walls with nails. Then prop up the two end roof frames where you place the end walls under.

9. Nail treated plywood to the top of the roof frames and make sure that the roof frames are vertical and parallel with each other.

10. Now, while making sure that your sidewalls are vertical, we can begin framing the sidewalls of your greenhouse.

11. Using your two-by-twos or two-by-fours, remember to frame in your doors on both sides with this plan. You will need to cut these

studs to size. They will be directly in the middle of the end wall and they will be a typical door size across, which is about 2.3 feet. Now, cut out the top board for the doorframe, which you measure for the proper width because of the sloping roof. Cut four more boards to fit between the sidewall and the doorframe.

12. At this point, put in your doorframe in the middle of each wall. On either side of the doorframe, frame in the smaller pieces that you cut to run from the sidewall to the doorframe. Once you have done this, cut two more boards that will run from the top corner, and connect the sidewall to the bottom corner connected to the doorframe. Do this for both sides of the door and for both end walls — these are your bracing boards.

13. Now, place vents on your roof in the spot where you have the vertical 1-foot board that attaches to the 6-foot long board above it and the 4-foot board under it.

14. You will need to build your own doors for this greenhouse, but they are relatively easy to make, thankfully. For the sides of the door, use two-by-twos that measure 5.8 feet high, but verify how high you need by measuring your doorframe. The top and bottom of the door will be 2.2 feet long and again measure the exact size so you know they will fit. Attach plywood gussets to secure all the boards in place in all four corners and in the middle where you will have another 2.2-foot board running across. Then, install the hinges and fit the door to your greenhouse doorframe.

15. Now, cover the entire greenhouse with polythene. Staple the polythene into place and cover all the doors as well.

Greenhouse Plan
7x10 A-Frame

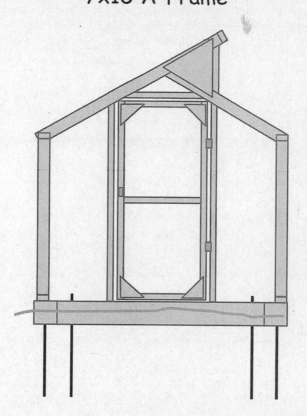

Bill of Material

Item	Qty	Description	Item	Qty	Description
1	1	8x8 timber x 6'-0"	11	1 roll	UV resistent poylethylene
2	2	8x8 timber x 10'-0"	12	6	1/2" plywood x 1'-3" x 1'-9"
3	10	1/2" dia. rebar x 3'-0" LG	13	3	2x2 x 2'-2"
4	5	2x4 x 10'-0"	14	2	2x2 x 5'-9"
5	6	2x4 x 6'-0"	15	4	2x4 x 5'-10"
6	6	2x4 x 4'-0"	16	4	1/2" plywood x 6" x 6"
7	12	2x4 x 4'-6"	17	2	hinges
8	6	2x4 x 1'-6"	18	1	handle
9	1 box	10d nails x 3" LG	19	1 box	#8 wood screws x 2 1/2" LG
10	20	1x4 x 1'-10 1/4"	20	1 box	#8 wood screws x 1 1/2" LG

Step 1: Assemble Timber Foundation:

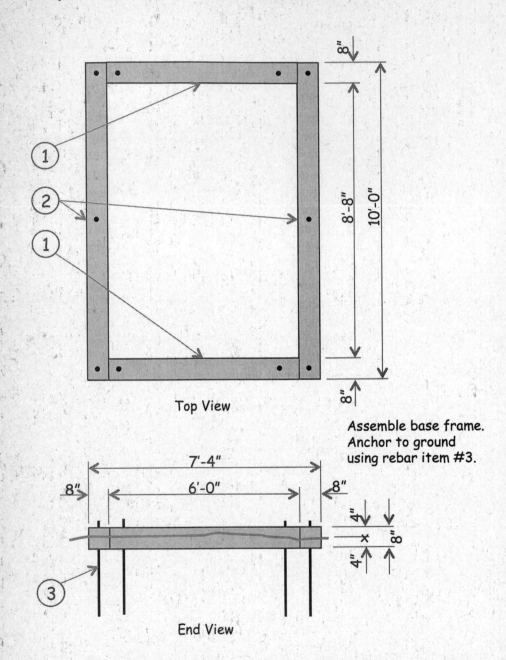

Top View

End View

Assemble base frame.
Anchor to ground
using rebar item #3.

Step 2: Assemble Framing:

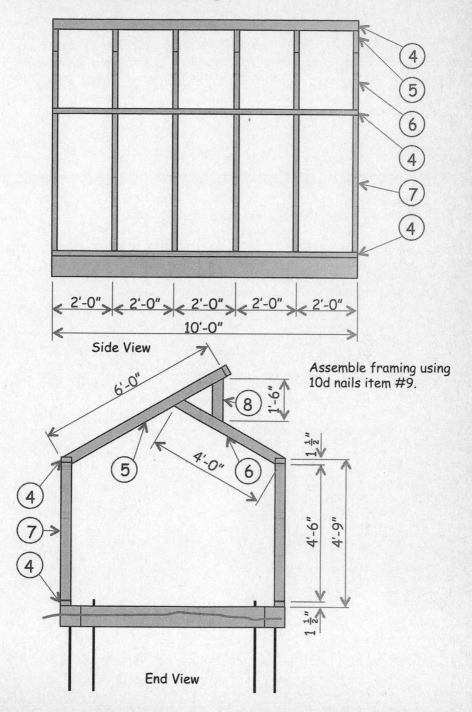

Side View

2'-0"   2'-0"   2'-0"   2'-0"   2'-0"

10'-0"

Assemble framing using 10d nails item #9.

6'-0"

1'-6"

4'-0"

$1\frac{1}{2}$"

4'-6"

4'-9"

$1\frac{1}{2}$"

End View

Step 3: Final Assembly:

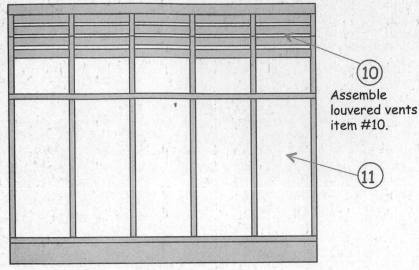

Assemble
louvered vents
item #10.

Side View

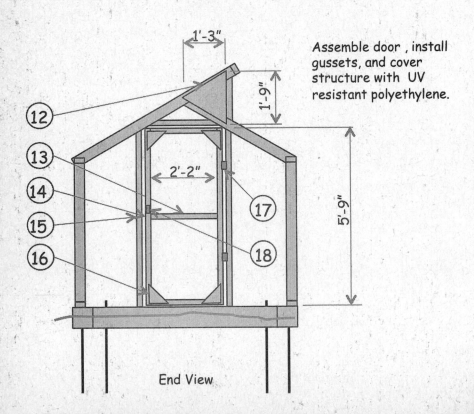

1'-3"

1'-9"

Assemble door , install
gussets, and cover
structure with UV
resistant polyethylene.

2'-2"

5'-9"

End View

A-Frame Greenhouse Plan No. 3

This A-frame greenhouse plan is for the expert greenhouse builder. Do not attempt this plan unless you have constructed something of this magnitude before. Although this greenhouse does not require the construction of doors, beginners should not try this plan unless guided by someone with experience.

1. Create a 10 x 10 base for your greenhouse. For the base, use 2 x 6 pressure-treated boards. The front boards will be 10 feet long and the sideboards will be 9 feet 1-inch. Under the base, reinforce the boards by installing gussets. Cut the gussets from one 45 x 16-inch board. Using 3/8 exterior-grade plywood, begin from the side and cut the following triangles (refer to diagram for approximate shapes).

    a. Cut one right-angle triangle that measures 16 inches on the one side (no cutting needed for this side as this is the edge) and 9 inches on the other side that is at a 90-degree angle from the 16-inch side. Do the same on the other side of the board, ensuring that the bottoms of the triangles are all 9 inches long.

    b. Cut the next two pieces so that the bottoms of the triangles are 18 inches long. Cut the one closest to (a) first and then you only need to make one cut for the other (b).

    c. Cut out the next four pieces so that each bottom of the triangle is 9 inches long.

Use (a) for the bottoms of the base. Depending on the size of your greenhouse, the length of these cuts will change.

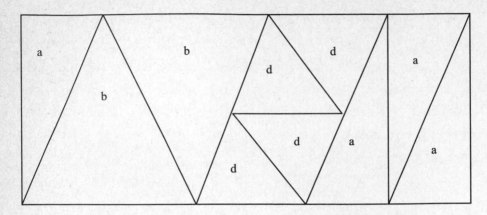

2. Now, use a screw fence anchor to anchor the greenhouse to the ground so it does not tip over when the weather is very windy or stormy.

3. You now need to make the rafters for the greenhouse. You will use two-by-fours that are 10 feet long. The rafters will connect at the proper angle (60 degrees) to the base of the greenhouse. At the top, each board will then connect to a one-by-four 10-foot long ridge board. Each side of the ridge board will have a rafter connecting to it. Nail each rafter into place every 2 feet, 6 inches. To secure it all together, use gusset (b) on either side of the rafters in order to secure the sides to each other and the ridge plate.

4. At this point, you need to make the end walls. From each side, place a stud that is 2 feet away from the sidewall and measure to fit with the proper angle.

5. Measure 1-foot and 9 inches from each stud and put in another stud. This will serve as the sides of your doorframe. The top and bottom of your doorframe will be two-by-fours measured to 2 feet 6 inches each (or whatever you need to fit them in). Above the top of the doorframe, install two vertical two-by-fours about 1-foot apart to help brace the roof. Measure out a space for the door and build the door so that it measures properly within the doorframe.

6. Cover the entire greenhouse and the door with plastic sheeting. You can also install a vent on the side of one of the end walls. The other end wall has the same design as the end wall with the door, but there just is no door.

Summary

A-frame greenhouses look great, but one of the biggest disadvantages to this type of greenhouse is the limited floor space in comparison to the overall height of the greenhouse. Another disadvantage is that the A-frame greenhouse is not perfect for the winter. During the winter, keeping heat inside the greenhouse is vitally important, and with an A-frame greenhouse you will have much of your heat rising above the plants and thereby leaving the plants relatively cold, which could damage their growing process. Usually, putting a fan at the top of the greenhouse can help redistribute the heat. And remember, with any of your greenhouse projects; be sure to wear protective eye gear if you are working with heavy machinery.

Chapter 7

Constructing Hoop Greenhouses

"How deeply seated in the human heart is the liking for gardens and gardening."

— Alexander Smith, poet

One of the cheapest forms of greenhouse to build is the hoop greenhouse. The hoop greenhouse is easy to build and very affordable and the perfect type of greenhouse for any gardener, from the hobby gardener to the commercial gardener. A hoop greenhouse is, put simply, a series of plastic hoops secured to the ground with some sort of covering over top of the greenhouse, like plastic. It is very simple to make because all you do is secure the hoops to the ground in intervals and then secure the plastic covering over the hoops.

The hoop greenhouse is exactly as it sounds, a series of plastic hoops usually made from PVC piping that you lay plastic over. It differs from an A-frame greenhouse in that instead of boards creating a roof, it has a series of half hoops joined together. This type of greenhouse is very popular, especially in areas that get plenty of rain, or have short growing seasons. The hoop greenhouse allows for the easy run off of rain and snow, it protects your plants from wind, and you can put a hoop greenhouse nearly anywhere you want on your yard. Hoop greenhouses also have very simple foundations to create.

There is no great amount of foundation preparation needed for a hoop greenhouse, because the hoops secure to stakes in the ground, not to the foundation.

The only main drawback of the hoop greenhouse is that the plastic lining that coats the entire greenhouse will need to be replaced every four to six years, on average. Other greenhouses do not need to have their covering replaced as much. However, it is easy to replace the covering because the hoop greenhouse is so easy to build, and the cost is very low. The structure of the greenhouse, however, is usually quite strong and will last for a long time.

There are many gardeners in northern latitudes who will grow vegetables in the winter through the use of a hoop greenhouse and a heater. Through the use of a hoop greenhouse, you can extend your growing season by as much as three to four months. For those living in northern latitudes, this means instead of having a growing season from May to August, they have a growing season from March to October. The temperature can easily be regulated within the greenhouse without any extra equipment beyond a heater in the winter. The hoop greenhouse will be warmed by the sun and cooled by the wind, which will keep temperatures inside pleasant. On nice days, you can even roll up the sides of the hoop house to allow a breeze in, or to get more sunlight in. This is done by removing the clips holding the plastic to the hoop piping and using ropes to roll the plastic up. This also helps to get some extra air flow into the greenhouse to freshen things up for the plants.

Unlike other improvements to your home, or even a large garden greenhouse, you do not need any special permits to build a hoop greenhouse because it is not considered to be a permanent structure on your property. It is not considered a permanent structure on your property, because it has no foundation.

Like A-frame greenhouses, there is plenty of space above you in a hoop greenhouse. This allows you to take advantage of that space by hanging baskets where you can plant things, like tomatoes. This is wasted space if you do not use it, and a line of plants going down the length of the hoop house can ensure you are getting the most bang for your space within a hoop greenhouse.

Probably one of the biggest benefits that come from the hoop house is the fact that you can move it, without taking apart the greenhouse. The walls of the greenhouse help to protect your plants not only from the elements, but from predators as well. Crop predators will be kept out with the hoop house serving as a barrier and that will increase the yields of what you grow.

Hoop Greenhouse Plan No. 1

As with any other kind of construction project, you must first measure off the area where you intend to place your greenhouse. With this first hoop-style greenhouse plan, you will need to make sure you have a large amount of space and an ample amount of time to spend on this project. If you do not have at least a 21 feet long and 10 feet wide space, do not attempt this plan. Before you begin to build your hoop greenhouse, you will need the following tools:

- (16) 30-inch long 1/2-inch thick heavy duty PVC pipe

- (16) 10-foot long 3/4-inch thick light duty PVC pipe

- (7) 3/4-inch light duty PVC pipe

- (2) 3/4-inch PVC tee connector (three-way)

- (6) 3/4-inch PVC cross connector (four-way)

- 20 x 25 sheet of clear plastic.

- (16) 1-inch black poly pipe, 8-inch long

1. Measure out a space on your property that is 21 feet long and 10 feet wide.

2. Every 3 feet on either side, mark a spot on the ground with a stake. This is where each tube is going to go that forms the support structure of the hoop greenhouse.

3. To make the two end hoops for the hoop house, you will use two 10-foot long PVC pipes that are joined together with the PVC tee connector.

4. Connect the other six hoops using the 10-foot PVC pipe but connect them to each other using the PVC cross connectors instead. You can connect these to each other with glue, or with a rubber mallet that ensures each piece is fit securely to the other.

5. Now, in the stakes you put in the ground, put the 30-inch long 1/2-inch thick heavy duty PVC pipe over top of the stakes and secure them into the ground. Put each hoop on its selected stake to form the skeleton of the hoop house.

6. Connect each hoop at the top of the hoop greenhouse ridge line using the 3/4-inch light duty PVC pipe.

7. Secure plastic over the frame by slitting the 8-inch long black poly pipe so that it can create clips that will hold the plastic in place.

Greenhouse Hoop Plan #1
10 FT x 21 FT

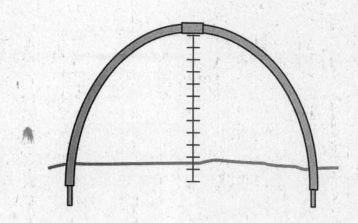

Bill of Material

Item	Qty	Description	Item	Qty	Description
1	16	1/2" PVC pipe x 3'-0"	5	1	Zipper
2	6	3/4" cross tee	6	1 roll	UV resistant polyethylene
3	16	3/4" PVC pipe x 10'-0"	7	48	3/4" PVC x 6"
4	2	3/4" tee	8	1 can	PVC cement

Step 1: Install Support Stakes:

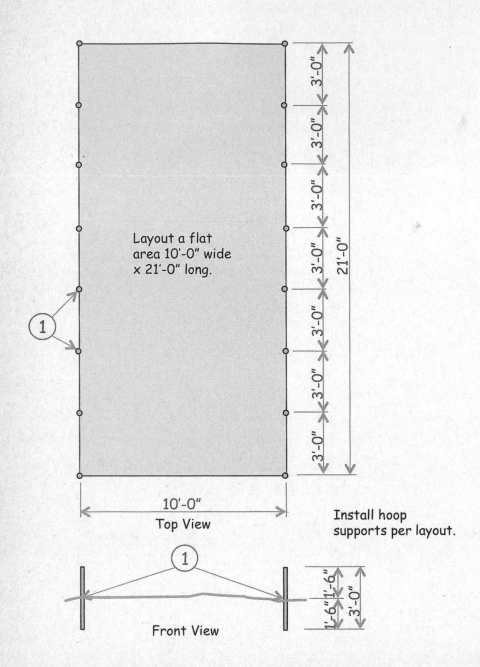

Layout a flat area 10'-0" wide x 21'-0" long.

3'-0"
3'-0"
3'-0"
3'-0"
3'-0"
3'-0"
3'-0"
21'-0"

10'-0"

Top View

Install hoop supports per layout.

1'-6" 1'-6"
3'-0"

Front View

Step 2: Install Hoops:

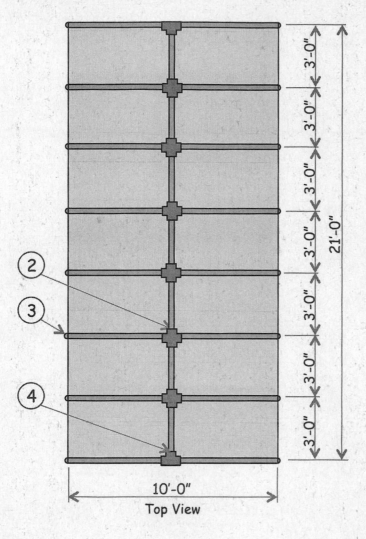

Install hoops into cross tee's.

Step 3: Final Assembly:

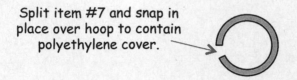

Split item #7 and snap in place over hoop to contain polyethylene cover.

Enlarged view of item #7.

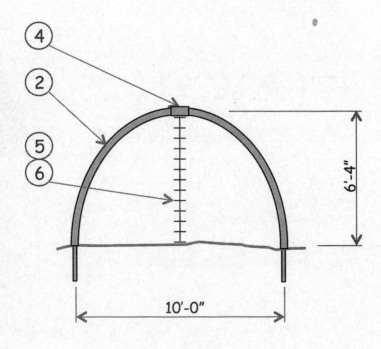

Front View

Hoop Greenhouse Plan No. 2

This hoop-style plan is designed for those without much building expertise. Beginners will find this greenhouse plan quick and easy to follow. To build this hoop greenhouse, you will need the following materials:

- 6 rigid PVC conduits

- 10 lengths of rebar

1. Measure out 15 x 4 feet for your greenhouse.

2. Pound rebar into each corner and then along the sides of the greenhouse every 3 feet with a sledgehammer. Make sure the rebar is far enough in the ground so that it cannot move very much. Typically, each piece or rebar should be 2 feet long and 3/8-inch thick.

3. Put one end of the PVC pipe over the rebar on one side and press down very firmly so the PVC pipe goes down into the dirt. It will be standing in the air now. The PVC pipe you should use is 1/2-inch in diameter and 10 feet long.

4. Grab the top end of the plastic hoop and bend it over to the other piece of rebar that is at the other end. Press this end firmly down into the ground just as you did on the other side.

5. Using plastic ties, adhesive, or clips, attach plastic sheeting over top of the greenhouse.

Greenhouse Hoop Plan #2
4 FT x 15 FT

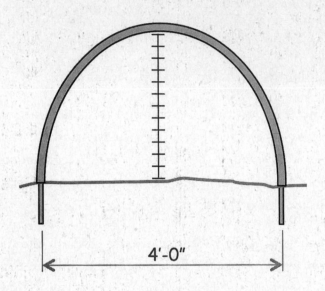

4'-0"

Bill of Material

Item	Qty	Description	Item	Qty	Description
1	12	1/2" dia. rebar x 3'-0" LG	4	1	zipper
2	6	3/4" dia. PVC x 10'-0" LG	5	1 pkg	plastic wire ties
3	1 roll	UV resistant polyethylene			

Step 1: Install Support Stakes:

Layout a flat
area 4'-0" wide
x 15'-0" long.

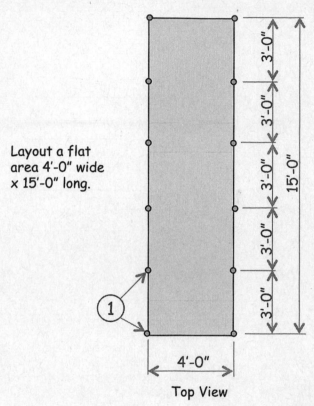

Top View

Install hoop
supports per layout.

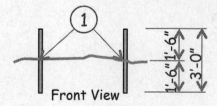

Front View

Step 2: Install Hoops:

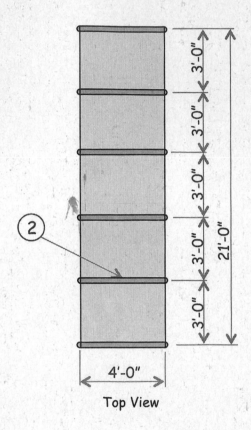

Top View

Install hoops over
rebar stakes.

Step 3: Final Assembly:

Cover PVC frame item #2
with UV resistant
polyethylene item #3.
Secure in place with
plastic wire ties item #4.

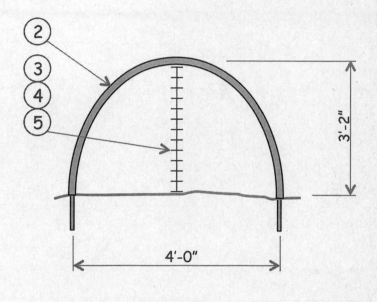

Enlarged
Front View

Hoop Greenhouse Plan No. 3

With this greenhouse plan, you should have someone assist you, given the amount of materials and time this project will require. This plan is also not specific to beginners or experts, but be sure you have help. In order to build this greenhouse you will need the following materials:

- (4) 2 x 6 timbers that are 16 feet long

- (2) 2 x 6 timbers that are 12 feet long

- (14) 2 x 4 timbers that are 12 feet long

- (19) 1/2-inch PVC pipe that is 20 feet long

- (9) 10 mm rebar that is 10 feet long

- (1) Bundle of staples

1. Lay down the two-by-six timbers in order to create a 12 x 32-foot frame. This means you have to join together two 16-foot pieces with a 2-foot piece of two-by-four. Take special care to ensure that your frame is completely square.

2. Cut each 10-foot piece of rebar into 30-inch pieces. You will then have 34 pieces that you need to pound 15 inches into the ground on the outside of the frame every 2 feet.

3. Take the PVC pipe and slid it over the rebar to make a hoop that spans the width of the greenhouse.

4. With metal banding that you can get from the grocery store, attach the PVC pipe to the two-by-sixes with screws.

5. Now you need to build the ends of the hoop house. So, the first thing you need to do is cut your 12-foot two-by-fours into the following sizes:

    a. (2) 11 foot, 8-3/4-inch pieces

    b. (4) 1-foot, 6-inch pieces

    c. (4) 4-foot, 7-inch pieces

    d. (4) 5-foot, 7-inch pieces

    e. (8) 1-foot, 11-1/4-inch pieces

    f. (2) 4-foot, 1/4-inch pieces

6. At this point, you need to construct the ends of the greenhouse by using the diagram displayed below:

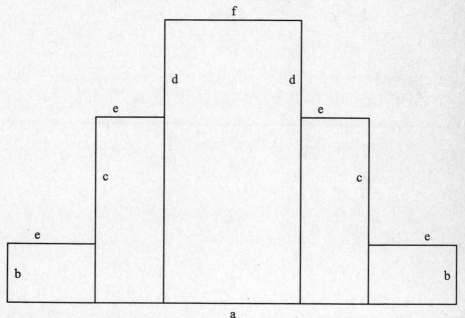

The layout looking at the front of the greenhouse.

7. Now, place the wall within the two-by-six frame, put it together, and brace it up with other pieces of wood that are secured to the base and secure the rest of the front and back structure to the PVC pipes using ties and metal banding.

8. Connect two PVC pipes, and then cut the PVC pipes so they measure 32 feet long total. This is the top of your hoop and the ridge to create the secure structure. Attach it to the rib structure using plastic zip ties.

9. You need to cover the entire structure in plastic now. Secure the plastic to the frame and ridges with staples. Put the plastic over the length of the greenhouse and make sure it overlaps on the ends to cover the end walls. Pull the plastic tight and attach it to the two-by-six boards at the ends and along the bottom. At the end where you are putting in a door, make sure you cut a space for the door and leave a few inches hanging over the door area so that you can wrap it inside and attach it to the doorframe.

10. You need to put in the door now. With the two-by-fours you have left, cut the boards into two 4-foot, 11-inch pieces and two 3-foot, 9-inch pieces. Nail these pieces together to create the door then cover this with plastic to create the door. At this point, put hinges on it and attach it to the doorframe.

Greenhouse Plan
12 FT x 32 FT

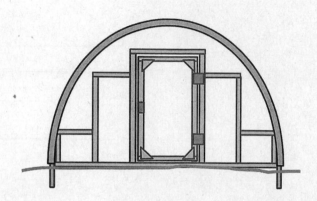

Bill of Material

Item	Qty	Description	Item	Qty	Description
1	34	1/2" rebar x 2'-6"	11	4	2x4 x 3'-8 1/2"
2	4	2x6 x 16'-0"	12	4	2x4 x 5'-3"
3	2	2x6 x 12'-0"	13	1 box	#8 screws x 1 1/2" LG
4	1 box	10d nails x 3" LG	14	34	PVC pipe x 10'-0"
5	4	2x4 x 1'-6"	15	34	PVC pipe couplings
6	8	2x4 x 2'-0"	16	1 box	wire ties
7	4	2x4 x 4'-7"	17	4	2" hinges
8	4	2x4 x 5'-7"	18	2	door handle
9	2	2x4 x 4'-0"	19	2	door latch
10	8	1/2" plywood x 6" x 6"	20	1 roll	UV resistant polyethylene

Step 1: Install Support Stakes:

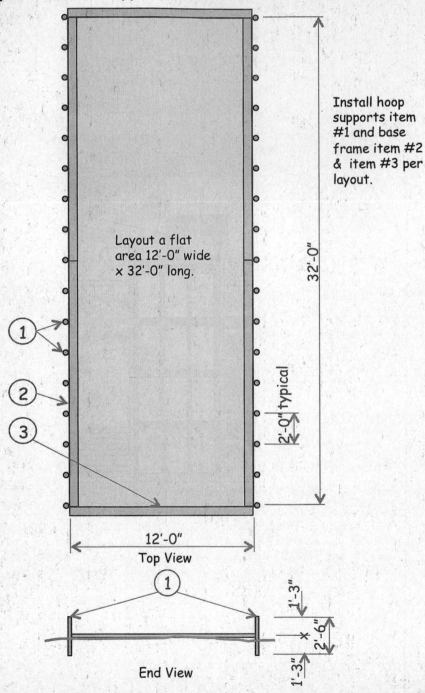

Install hoop supports item #1 and base frame item #2 & item #3 per layout.

Layout a flat area 12'-0" wide x 32'-0" long.

32'-0"

2'-0" typical

1

2

3

12'-0"

Top View

1

1'-3"

x

2'-6"

1'-3"

End View

Step 2: Assemble End Frames:

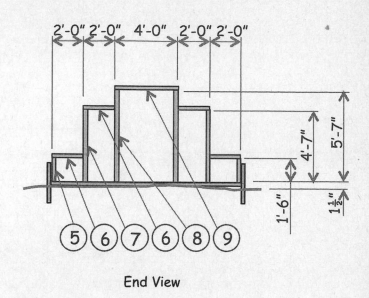

End View

Step 3: Assemble Doors:

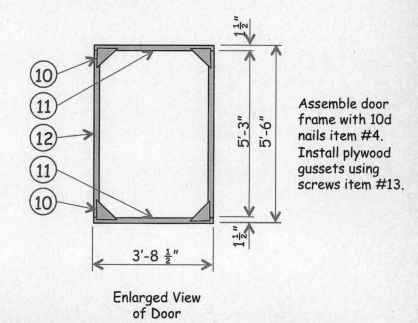

Assemble door frame with 10d nails item #4. Install plywood gussets using screws item #13.

Enlarged View
of Door

Step 4: Final Assembly:

Attach PVC pipe item #14 end to end using PVC couplings item #15. Place PVC pipe over rebar and secure with wire ties item #16. Install door assembly using hinges item #17, latch item #18, and handle item #19. Cover entire structure with UV resistant polyethylene item #20.

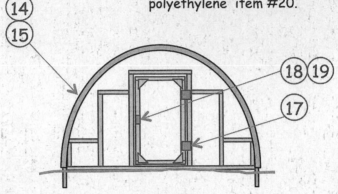

End View

Hoop Greenhouse Plan No. 4

Although this hoop-style greenhouse may potentially be more time consuming than the other plans, it requires no machinery to construct. The PVC piping is durable enough that it will bend without breaking, so this is the perfect do-it-yourself project. To make this greenhouse, you will need the following materials:

- 213 feet of 1-inch thick PVC pipe

- 40 inches of 1-1/4-inch PVC pipe

- (30) 1-inch slip Ts

- (4) 1-inch three-way connectors

- (8) 1-inch 90-degree elbows

- (10) 12-inch ground stakes

- (130) 1-1/4-inch snap clamps four inches long

- (13) 3/4-inch self tapping screws

- 20 x 25 feet plastic sheeting

1. Measure out and cut the 1-inch thick PVC pipe to the following measurements:

   a. (2) 11-1/2 feet pieces (front and back pieces)

   b. (2) 12-foot, 3/8-inch pieces (side)

   c. (5) 19 feet long pieces (arch)

   d. (4) 3 feet long pieces (spine)

    e.   (4) 6 feet long pieces (side struts)

    f.   (2) 67-inch long pieces (door sides)

    g.   (10) 27-inch long pieces (door/window struts)

2. You should also cut the 1-1/4-inch pipe into ten 4-inch pieces.

3. Cut the plastic into the following:

    a.   20 x 13-foot, 4-inch piece (top sheet)

    b.   (2) 12-foot, 1/2-inch by 7 feet pieces (end pieces)

    c.   6-foot, 6-inch by 3-foot, 4-inch piece (door)

    d.   3-foot, 4-inch square (window)

4. Bend each arch piece to form an 11-1/2-foot wide U-shape.

5. Glue a 1-1/4-inch thick piece to one of the snap clamps to form a figure eight. Do this twice.

6. Glue 1-1/4-inch thick pieces to each other to form a figure eight in order to make hinges. Do this eight times.

7. Use a slip T and a hinge that you have just made onto a door side and attach a 90-degree elbow connector at the top and the bottom. Connect three door struts to the elbows and attach the other door side with two elbows and one slip T to make the two-panel door. Glue all these pieces, but not the hinges, together.

8. With the four window pieces and the four elbow joints, put together a square window and put two hinge pieces on the same side before you glue the elbows to the window pieces.

9. Put three slip Ts onto the side pipe pieces and then put two slip Ts on the front and back pipe pieces. At this point, you are not gluing anything.

10. With a three-way connector, assemble a rectangular base frame out of the front, back, and side pieces.

11. Attach the base frame to the five stakes on each side that you have nailed into the ground.

12. Put two arches that will be the front and the back, and put three slip Ts in the middle of these. On the remaining three slip Ts, put in the two slip Ts.

13. Put the front arch onto the three-way connectors at the corner of the base frame you have created and do the same for the rear arch.

14. Place one of the remaining arches onto the two slip Ts that are nearest the front archway and then attach the middle slip T on the archway to a slip T on the second arch way with one of the spine pipes you have created.

15. Do this for the remaining arches.

16. Make the door so it looks like an H by slipping a slip T to each onto the side struts and then attaching those to the door strut. Slid the door onto one side of the frame and put the latch on the other side. Then, glue together the H-frame, latch, and the hinges.

17. Make a second H-frame and attach the last latch to the middle. Put the window hinges on the window strut and attach slip Ts to both sides so that you can slide it onto the bottom. Glue the hinges, frame, and latch together.

18. Put the window and door onto the slip Ts that are hanging. The door should be at the front and the window will be in the back.

19. Now, you will put in the covering. With snap clamps, clip the bottom of the front of the plastic sheet to the front base piece and fold the corner that is loose over the arch. Be sure to mark out where the door is and then cut the door out. Clamp the plastic down so that you are using clamps every 8 inches. Continue to do this until you have reached the back of the greenhouse, and then cut a space for the window.

20. Cover the window and the door with plastic and clamp them 10 inches apart.

21. Trim any excess plastic.

Heating Your Hoop Greenhouse

One of the most important things to do with your greenhouse is heat it when the weather gets cold. This helps you have a greenhouse that stays warm and it gives you a longer growing season. With some heating within your greenhouse, you can ensure you have a growing season that is two to three months longer. In addition, there is also the need to provide some sort of heat to your greenhouse at night. During the early part and late part of the growing season, nights will be colder and you are always at risk for freezing temperatures. In order to do the following, you will need these supplies:

- **Shovel** – You will use this to dig the trenches.

- **Manure** – This will be poured onto the wood chips.

- **Wood chips** – The trenches will be filled with these wood chips.

- **Water barrels that are black** – You will use the water barrels to absorb the heat inside your greenhouse.

- **3/4-inch plywood boards** – These will be used to cover the trenches after the wood chips are placed inside.

- **Planting flats**

Perform the following steps to ensure your greenhouse has warmth at night and during the late fall, winter, and early spring months.

1. Dig three trenches in rows along your hoop greenhouse. These trenches need to be 3 feet wide and 3 feet deep. Ensure that you have 1-1/2-feet of space between each of your trenches.

2. With the trench in the middle, fill it halfway with wood chips.

3. Pour manure onto the wood chips throughout this trench and mix it up. There should be two parts wood chips to one part manure. The breaking down of the manure and wood chips will actually generate heat to help warm the greenhouse during the evening.

4. Put plywood over the center trench so that you can walk over it.

5. On the two side trenches, place the black rain barrels along their sides. Fill each black rain barrel with water and then put plywood boards on top of the barrels. This will allow you to have planting tables you can use. During the day, the heat inside the greenhouse will be absorbed into the barrels and that will heat up the water during the day. The heat released at night will help warm the greenhouse.

6. Every few days, take the boards off the center trench and mix up the wood chips and manure. Once composed after a few months, you can use it in your garden as fertilizer.

Summary

The hoop greenhouse is a very popular type of greenhouse because of how easy it is to build. With PVC piping, plastic sheeting, some clips, and a bit of knowledge, you can build an excellent hoop greenhouse. Hoop greenhouses are portable and inexpensive as well. Because they are not permanent buildings on your property, you do not have to worry about building codes, and one day on the weekend is more than enough time to build a hoop greenhouse. Along with lean-to greenhouses, hoop greenhouses are perfect for beginners and those who are not as experience building a greenhouse by themselves.

Chapter 8

Cold Frame Greenhouses

> *"Gardens are not made by singing 'Oh, how beautiful,' and sitting in the shade."*
>
> — Rudyard Kipling, British Author

The cold frame greenhouse is small, easy to build, and you can move it anywhere you need. It is a great option if you do not have much space, or you are not growing many vegetables. A cold frame greenhouse is a transparent, roofed enclosure that is built low to the ground with the purpose of protecting plants from the frigid weather that comes in the fall and spring. With the transparent top, sunlight comes in but heat does not escape, keeping the inside of the cold frame greenhouse quite warm.

In the past, cold frames were built as an addition to a heated greenhouse. However, these days — in northern latitudes — the cold frame greenhouse serves as a great addition to the outside of a home as a way to keep growing vegetables in cold weather much longer.

Cold frames are found in many different places, including large scale farming and home gardens, because of their ability to create microclimates. A microclimate is a small eco-system created in a localized space where the climate inside is different than the surrounding climate. A microclimate occurs because the air and soil temperature is hotter inside the cold frame, and the cold frame shelters your plants from the wind. With a cold frame greenhouse, you can grow plants into the winter and much earlier in the spring. Many gardeners will use cold frame greenhouses to grow seedlings that are then transplanted out into the open ground, or to grow plants that are resistant to the cold.

Traditionally, a cold frame greenhouse will use glass windows, a wood frame, and will be built to about 1 to 2 feet high. The window will be placed on top of the greenhouse and the roof will be sloped in a manner that allows for the most winter sun to get into the box and stay in, thereby keeping it warm. Some builders will use clear plastic or sheeting instead of glass.

This is a cold frame that is attached to the outside of a free-standing greenhouse.

Typically you can build a cold frame greenhouse for less than $100. Depending on the size of your cold frame greenhouse, however, you will pay more. Using glass on the top of your cold frame greenhouse is going to drive up the cost more so than if you were to use plastic.

Things to Consider

When you are putting together your cold frame greenhouse, there are some things to consider before you start building.

Location

The location of the cold frame greenhouse is highly important. If you put the cold frame greenhouse in a spot that is shaded, you will be robbing your greenhouse of much needed sunlight. You want to have your greenhouse facing to the south where it can get the most sunlight during the day. This is especially true during the winter when sunlight is not as plentiful, but is vitally important.

Support

The next thing to consider is supporting your cold frame. What are you going to put the cold frame greenhouse against? It needs support, typically, from a home, garage, or solid fence. Although you can have a cold frame greenhouse that is standing alone, the building you put the greenhouse against helps greatly. For example — using a home — the heat from the home will warm the cold frame greenhouse, thereby warming the inside of the box. Placing the greenhouse in the right spot — and on the right side of the house — can also protect it from wind and the elements, which allows it to last longer.

Drainage

Drainage is also important with your cold frame greenhouse. If your plants are sitting in water because you put the cold frame greenhouse under the eaves, or if you do not have a good enough slope in the box, you will not be helping yourself. This is why it is not only important to make sure that the greenhouse can drain water, but that it does not take on too much water either.

Advantages of Cold Frame Greenhouse

As can be expected, these small greenhouses are truly great and have plenty of advantages associated with them. Some of the benefits of cold frame greenhouses include an early and longer growing season, aesthetics, and pest control.

Early growing season

You can start your seedlings earlier in the dirt without worrying about the cold. Many gardeners have to start their seedlings inside, which can take up room and be a bit dirty. With a cold frame greenhouse, you do not have to worry about that and you can start your seedlings nice and early. Make sure you check the temperature of the greenhouse before you start planting anything because you want to make sure that the soil is warm enough to promote the growth of the plants in this greenhouse. Remember, you want the temperature to be between 50 and 85 degrees.

Longer growing season

The cold frame greenhouse gives you longer growing seasons and that allows you to not only grow more, but to grow different varieties of vegetables. You will be able to grow different types of vegetables, flowers, and plants with the use of the greenhouse. You can even use this to grow herbs well into the winter, which you can then use in your home.

Aesthetics

When you build a really nice cold frame greenhouse, you actually help make your home look nicer from the outside, which can even increase the overall value of your home. By decorating the cold frame greenhouse, and designing it so it adds to the exterior of your home, instead of standing out, it can become something quite attractive on your property.

Pest control

Slug control is a big part of your cold frame greenhouse. With your cold frame greenhouse, you can keep slugs out and therefore protect your vegetables and fruits from destruction. You do not want to ruin your harvest, so you need something to control the slugs. With a cold frame greenhouse, you keep out slugs, beetles, pests, and snails.

How to Use a Cold Frame Greenhouse

Before you learn how to build your greenhouse, we will cover how to use a cold frame greenhouse so that you can get the most benefits out of it. You want to get the maximum yield from your cold frame greenhouse and knowing how to use it properly is a big part of that. If you are going to use a cold frame greenhouse, you should do the following:

1. Adjust the light coming into the structure to maximize the amount of sunlight that reaches the inside of the cold frame. You do not want any parts being shadowed within the greenhouse because this will rob many plants of important sunlight.

2. The cold frame greenhouse is a garden, just a different type of one, so when you are gardening in it, garden as you would in the garden. You do not have to take any special considerations when you are using a cold frame greenhouse.

3. Keep an eye on the temperature inside your cold frame greenhouse. Depending where you are, the temperature outside can change greatly, by as much as 10 or more degrees in a day. This means that the cold frame greenhouse may be at the right temperature at 11

a.m. in the summer, but by 3 p.m. the heat inside the cold frame greenhouse could be so hot that it kills the plants, especially if it goes above 80 degrees. Typically cold frame greenhouses do not have any ventilation, so there is nowhere for the heat to go inside this greenhouse. During hot days, prop the top of the greenhouse up a bit to allow cool air to come in and to help stabilize the temperature.

4. The soil moisture is very important because in cold frame greenhouses it can be very dry. You want to make sure your soil within the greenhouse is moist — no different than you would in the garden itself.

Cold Frame Greenhouse Plan No. 1

Cold frame greenhouses are very simple to make, and this is pretty much the simplest type of cold frame greenhouse that you can construct. This plan is not specifically for beginners, but it is an easy plan to follow if you do not have much experience building greenhouses. The supplies you need include the following:

- Cardboard box

- Utility knife

- Duct tape

- Plastic sheeting

1. Find a box that is large enough for your plants. To maximize the efficiency of the heat within the greenhouse, you want to make sure that you do not have much wasted space. Therefore, only select a box that is slightly larger than the space your plants need. You should have a few inches of foliage between the plants and along the sides of the box.

2. Cut the flaps off the cardboard box.

3. Cut plastic sheeting to slightly larger than the size of the opening of the box. You may need a few pieces of plastic wrap depending on the size of the box.

4. Place the plastic over one of the ends of the box and secure it with duct tape.

5. Place the plants within the cold frame greenhouse and put them outside where they will get the most sunlight. Another option you can use, especially when the temperature is getting colder in the fall, is to cut the bottom off the box and go out to the garden with it and place it on the plants you want to protect. If you have these all ready, you can place them all over your plants throughout the garden to protect them during the cold evenings. If you are just putting the box over the plants, make sure that you push the edges of the cardboard cold frame box deep into the ground to offer the most protection. If it is windy, you do not want to lose the box and leave your plants unprotected against the elements.

Cold Frame Greenhouse Plan No. 2

This cold frame greenhouse is a bit more complicated than the other ones, but still relatively easy to build. This plan is reserved for someone who has more experience building, but people who considers themselves beginners may also want to give this a try. Just keep in mind that with any greenhouse plan you choose, you will need to give it the time it deserves. You will need the following materials:

-   Square bales of hay

-   Clear plastic sheeting

-   Stones

- Bricks/mortar

- Lumber two-by-fours

- Window frame and glass or roofing plastic

- Hinges

1. Select a size for your cold frame greenhouse. It should be about 3 x 6 feet and make sure you put it where the sun will hit it the most. Build the cold frame greenhouse so that the side that faces to the north measures about 18 inches tall, while the other side is only 12 inches tall. This is to create the proper slope for the sun to enter into the greenhouse.

2. Prepare the soil within the frame of the greenhouse in the plot you are building it on. You want it to be ready to plant while you can still work the soil easily, because the walls are not in yet.

3. Build the frame around the soil you have worked on. You can build it with bricks and mortar or you can use two-by-fours. Use whichever foundation you are more comfortable with using.

4. On top of the frame, attach an old window sash to the north side of the greenhouse using hinges and attach the window to the hinges.

Cold Frame Greenhouse Plan No. 2.5

Another option if you are building a cold frame greenhouse is to use hay bales. You can build this cold frame by doing the following.

1. Build the cold frame walls out of hay bales.

2. Put thick plastic sheeting over the bales and secure them under bales or use heavy stones/bricks to keep the plastic tight.

3. Make sure that the plastic does not touch any plants.

Cold Frame Greenhouse Plan No. 3

If you can make a wooden frame and have a spare window, you can try using this cold frame greenhouse plan. Much like the previous plan, the more experienced builder will want to consider this project. Beginners should not be deterred, however. Keep in mind that it is recommended you have someone working with you who is knowledgeable about construction, especially someone familiar with installing a thermometer. You will need the following materials for construction:

- Window sash

- Wooden frame

- Corner braces

- Thermometer

- Hinges

1. When you are putting together the four sides of your cold frame greenhouse, you need to choose the right wood for the frame. The best wood to choose is cedar or cypress, as it will not decompose as quickly.

2. Cut the wood out so that its measurements equal the measurements of the window you are putting in. You want to ensure that the window fits snugly on top of the frame so that no heat is escaping, and no cold air is coming in. Once you have cut out the pieces of wood. Attach them to each other using braces. Do not forget to ensure that the back of the frame is higher than the front of the frame. You want to get a good angle for your cold frame greenhouse so you let in the maximum amount of light and heat from the sun.

3. Put heavy-duty hinges on the window, because these will keep it shut on windy days. Put the thermometer in the greenhouse and keep an eye on it. You want to keep the temperature in the greenhouse at about 70 to 75 degrees during the daytime and 55 to 60 degrees during the night.

Cold Frame Greenhouse Plan No. 4

This plan only requires that you know how to cut. This plan should be considered by beginners. This should only take you an hour or two to complete because of the sheer ease of this type of greenhouse. You do not need to plan out an area because this greenhouse is moveable and can be placed anywhere for your convenience. For this cold frame plan, you will need the following supplies:

- Piece of glass the same size as the box

- Wooden box

- Woodscrews (large and small)

- O-rings (rubber) – These are like washers — to help create an airtight seal

- One piece of wood the same thickness as the glass

1. Cut out the bottom of the wooden box. You can throw this in the compost heap, throw it in the garbage, or burn it because you do not need it anymore.

2. Fit one side of the box with two screws and two rubber rings. The glass is going to rest on these rings and you only need to have one side of the box with these rings.

3. Mount the glass on the bottomless box by attaching it to the piece of wood sitting upright and attach that piece of wood to the wooden box.

4. Place the wooden box with window on it over the plants you want to put in the greenhouse.

Cold Frame Greenhouse Plan No. 5

This cold frame plan does not require any kind of heavy machinery use, but keep in mind that this is still a construction project — albeit a small one. Before you finish this greenhouse, you will have to pre-treat the wood you use with a preservative to protect it over time, so be sure you have gloves and goggles to protect your skin and eyes from harmful fumes. For this greenhouse, you will need the following supplies:

- Wood

- Window

- Screws

1. Cut two pieces of 2 x 1 wood so they are 6 inches high. These are for the front, internal corners. Cut two pieces that are 11 inches high; these are for the rear of the cold frame greenhouse.

2. At this point, screw the base together by fitting together the pieces that you just cut. Use wood glue between the sides to get some added strength along with the nails that you used. Place the angled top pieces on top of the greenhouse.

3. Paint the wood with a wood preservative on the outside to protect it from the elements.

4. Install the window on top of the cold frame greenhouse.

If you want to insulate this cold frame greenhouse, then you can put polystyrene tiles on the inside. This will hold in the heat very well and the white color of the tiles will reflect heat back into the box, rather than letting the sides absorb it.

Cold Frame Greenhouse Plan No. 6

This greenhouse plan will require more time and care to complete. You will need to follow the instructions carefully, as this greenhouse calls for many building materials. Although it is still a small scale greenhouse, be sure to allot yourself two to three days for completion. For this cold frame greenhouse, you will need the following supplies:

- 4 x 8 feet of 3/4-inch plywood

- Corner braces

- Plastic liner

- Shovel

- 26 x 36-inch fiberglass panel

- 8 feet of two-by-two treated lumber

- Wood screws

1. Cut the sides, front, and back panels to the following dimensions, along with corner braces: (all values in inches)

    a. Back Panel: 26 x 36

    b. Front Panel: 22 x 36

    c. Side panel: 22 x 26 x 28-1/2

    d. Vertical cover frame: 5-1/2 x 30

    e. Horizontal cover frame: 4 x 36

    f. Back corner braces: 26

2. Put the box together by attaching the side, front, and back panels to the front and back corner braces.

3. Using a saw, cut a 3/4-deep saw gap down the center of each horizontal cover frame.

4. Drill holes into the cover frame to use as pilot holes and then attach the fiberglass into the kerfs with wood screws.

5. Set the cover frame on top of the vertical frame pieces and fasten them together.

6. Fasten the fiberglass panel to the vertical frame pieces.

7. Attach the butt hinges to the inside of the cover, roughly 8 inches away from the back corner.

8. Completely align the cover with the edges and attach it to the hinges.

9. Dig a hole that is 36 x 42 x 24 inches deep, put 1-foot of manure in it, and then put the cold frame in this hole.

10. Put dirt around the edges of the cold frame to secure it in.

Cold Frame Greenhouse Plan No. 6

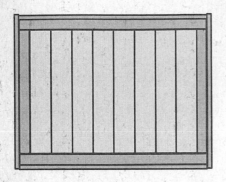

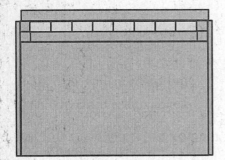

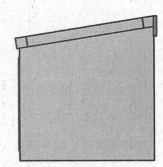

Bill of Material

Item	Qty	Description	Item	Qty	Description
1	4	2x2 x 3'-0"	8	1	3/4" plywood x 2'-2" x 3'-0"
2	2	2x2 x 1'-7"	9	1	3/4" plywood x 1'-10" x 3'-0"
3	2	2x2 x 2'-0"	10	1 box	#8 wood screws x 1 1/2" LG
4	2	2x2 x 2'-0 3/8"	11	2	2x2 x 3'-0"
5	2	2x2 x 1'-11"	12	2	2x2 x 2'-0 1/2"
6	1 box	#8 wood screws x 2 1/2"LG	13	2	hinges
7	2	3/4" plywood x 2'-2" x 2'-4 1/2"			

Step 1: Assemble Frame:

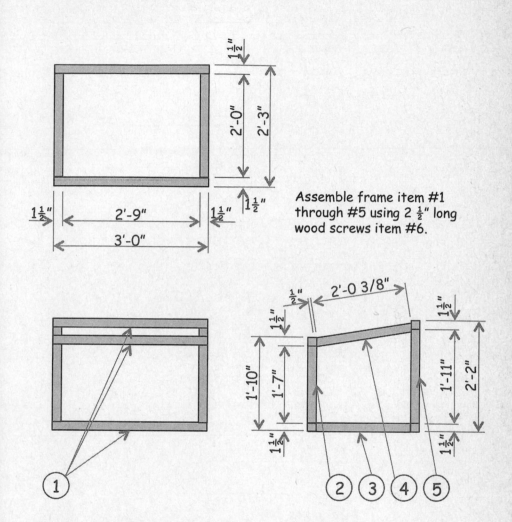

Assemble frame item #1 through #5 using 2 ½" long wood screws item #6.

Step 2: Cutout Plywood Panels:

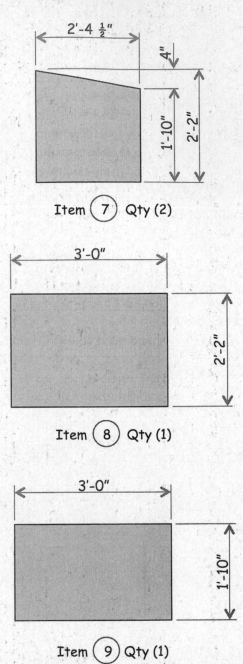

Item ⑦ Qty (2)

Item ⑧ Qty (1)

Item ⑨ Qty (1)

Step 3: Attach Side Panels:

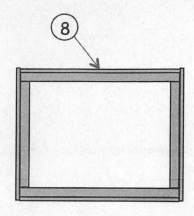

Attach side panels using 1 ½"
long wood screws item #10.

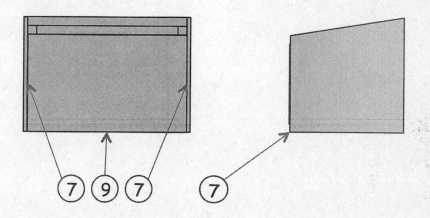

Step 4: Assemble Top Frame:

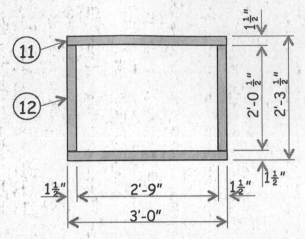

Assemble top frame and fiberglass cover using $2\frac{1}{2}''$ screws item #6.

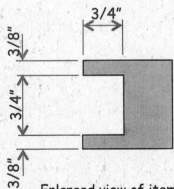

Enlarged view of item #11 & item #12. Place fiberglass panel in groove prior to final assembly.

Step 5: Attach Lid & Hinges:

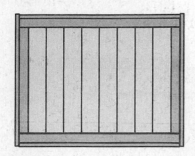

Attach lid using 2
hinges item #13.

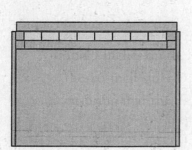

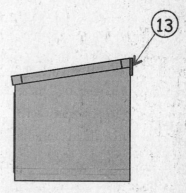

Cold Frame Greenhouse Plan No. 7

For this cold frame plan, you will first need to mark out the area in which the intended structure should stand. This is a relatively quick and simple plan to follow, so construction should only take a few hours to a day to complete. There are two different options for covering you can choose from with this cold frame — fleece or glass. Instructions for glass paneling immediately follow this plan. For this cold frame greenhouse, you will need the following materials:

- (2) 12-foot 2 x 8s or 2 x 10s

- (2) 4-foot 2 x 8s/2 x 10s

- (4) 2-foot 2 x 2s

- (7) 7 feet of 1/2 inch PVC pipe

- 8 x 16 feet of spun fleece

- Glass panes smaller than 18 x 30 inches with round edges

1. Mark out the area that you are going to be putting the cold frame greenhouse. Hammer in four-by-four posts into the ground to serve as the corners.

2. Nail the two-by-eights or two-by-tens against the posts to create the frame.

3. Put in metal brackets every 2 feet along the sides of the greenhouse for a place to position the PVC pipes.

4. Insert each PVC pipe into its bracket, bend it over, and insert it into the opposite bracket.

5. During the summer, cover the pipes with fleece to keep out insects and during the fall, put in plastic to keep the warmth in.

If you want to use glass instead of fleece or plastic, then you will need to do the following:

1. Position the glass so that each piece forms a triangle with the piece on the other side. You will be positioning the glass within the frame, so the two-by-eights or two-by-tens will keep the glass in position. You will probably need about three panes of glass for each side.

2. Clip each piece of glass to the other piece of glass using cloche clips. You can also make your own by cutting grooves into pieces of a two-by-four so that the grooves are at an angle for the glass to slide into, but the grooves of each side do not meet each other.

Cold Frame Greenhouse Plan No. 8

Much like the previous cold frame plans, this greenhouse is not specified for experts or beginners. Because of the sheer ease with which you can construct this greenhouse, anyone can attempt to build it. As with any greenhouse plan where you are using lumber, make sure that it is pre-treated with a protective coating so that it holds up against the elements. If you do not buy lumber that is already treated, it is recommended that you treat it yourself. For this cold frame plan you will need the following supplies:

- (3) 1 x 12s lumber 8 feet

- (2) 1 x 3s lumber 12 feet

- Plastic

1. Cut one piece of 1 x 12 in half so that each piece is 4 feet long.

2. Cut the second piece of 1 x 12 so that both pieces are 4 feet long. With the second 4-foot long piece, cut it diagonally into two equal triangular pieces.

3. Attach the pieces together. The three pieces that are 4 feet long will be the front and two pieces for the back. The two triangle pieces will serve as the sides.

4. Use brackets and nails to secure the pieces together. Remember to secure two 4-foot pieces on top of each other to create the higher back.

5. Dig a pit that is slightly smaller than the cold frame greenhouse box. Dig it deep enough though that you can fit in 10 inches of manure

6. Pour in 5 inches of manure, and then put a window screen on top. Put the box over this and then pour another 5 inches of manure in.

Summary

Cold frame greenhouses are simple greenhouses that are small, portable, and easy to build. If you live in an apartment building, have little space, or are new to greenhouses, then this may be the best greenhouse you can build for now. It can serve as the foundation for larger greenhouses that you build down the road. This type of greenhouse is also a great greenhouse for kids because of how easy it is to use and build.

Chapter 9

Building Accessories for Your Greenhouse

> *"We learn from our gardens to deal with the most urgent question of the time: How much is enough?"*
>
> — Wendell Berry, American poet and writer

Your greenhouse is only as good as the items within it. You have plants in there, of course, but you need to have accessories within your greenhouse, because this will make the entire process of using your greenhouse much easier. Some of these items include benches, potting tables, shelving units, growing beds, and more.

Potting Benches

Your potting bench is a workbench for use with gardening tasks that includes transplanting seedlings and caring for plants. A basic potting bench will have a work surface that is at a height where you can stand and work on it, while also offering you an area to store potting soil, pots, and tools.

Building a bench for your greenhouse is not the same as building a bench that you have inside your home. Your garden greenhouse benches will be exposed to water, soil, and sunlight and therefore must be capable of withstanding the elements without degrading too quickly. Wood like cedar is very popular, because it is hardy. Some gardeners choose to use metal for their benches but metal can corrode easily in the moist environment of a greenhouse.

Other gardeners will build potting benches that are small and can move easily. Some gardeners will also choose to attach the potting bench directly to the greenhouse itself. For those gardeners who go the extra mile, a dry sink for soil storage, a water drain, and a cold frame greenhouse box are also included on the bench.

Types of Potting Benches

Your potting bench is your workbench. This is where you do the tasks of the greenhouse and gardening. It makes things easier for you and it helps

Cedar potting bench.
Photo courtesy of Juliana America, LLC.

make your gardening much more efficient. It is the place to store things, make things, and nurture things. It is your greenhouse central station and if you love gardening, one of your favorite places to be. There are three main types of gardening benches to choose from:

Wood potting bench

This is probably the easiest to make but, because of the amount of moisture in a greenhouse, as well as the dirt, it can be easy for your potting bench to break down if you do not treat it properly. When you use cedar, you are using a wood that is naturally resistant to decay and warping, and it is very cost-effective as well. One thing to keep in mind is fertilizer salts on the bench. This salt can get into the wood, which will cause the wood to leach out oils that can get into the soil and cause harm to the plants growing near the gardening bench. Traditionally, potting benches are made out of wood because of how easy it is to work with wood. If you are worried about salt eating the wood, if it is not treated and is not cedar, or you worry about decay and warping, then you should consider a hardier material like metal.

Metal potting bench

These benches stand up to much more than wood benches, but they do cost more and are harder to put together unless you have experience working with metal. The moisture in the air of the greenhouse can cause the metal

Metal potting bench.
Photo courtesy of Juliana America, LLC.

potting bench to degrade faster as well. Metal potting benches are easy to clean, but they can be very heavy and difficult to move. Even though metal potting benches do decay in the humid air of a greenhouse, they will last longer than wood potting benches. They are a bit heavier to move around, and a bit more difficult to work with if you are building one, but they are highly durable.

Plastic potting bench

These benches are low-maintenance, portable, and lightweight. They are excellent choices if you have a bench near your plants, because they will not leach chemicals into the soil, nor will plastic benches react to fertilizer salts. The only problem is that plastic gardening benches are very susceptible to the elements, especially the sun. Over time, the sun will degrade the plastic of the potting bench, which will require you to replace it. Plastic potting benches are mass-produced so they are extremely cheap, often costing less than $75 on average. However, you will replace your plastic potting bench at a much faster rate (every three years) than you would wood or metal benches.

Potting Bench Plan No. 1

An experienced woodworker can finish this plan in a day, but if you are a beginner, allow yourself the weekend to complete this project. Also, be sure that you take safety measures when using the machinery needed to construct your potting bench. To build this potting bench, you will need the following tools:

- (3) 2 x 4s 12 feet long

- (5) 1 x 4s 8 feet long

- Assorted tools (Hammer, nails, saw, level)

1. Cut the following length of your two-by-fours:

   a. Eight pieces 3 feet long

   b. Three pieces 4 feet long

2. Cut all the one-by-fours into 42-inch lengths to create ten 3-1/2-foot long boards.

3. Put down two of the 3-foot boards you have made so that they are parallel to each other, and then lay one across the ends of these two boards. Make sure this cross beam is flush with the outside

edges of the two other parallel boards, which will serves as the legs of the bench.

4. Nail the two boards to the two pieces of wood securely. Place the fourth board halfway down the length of the legs and nail it securely in as well.

5. Do this again to recreate the second set of legs and braces.

6. Nail one of the 4-foot long two-by-fours to the outside edge of one set of the legs. The legs need to be facing out with the braces facing in. Nail the other end of the 4-foot board to the other set of legs. With this set of legs, make sure the legs face out while the crossbar braces face in. Once you have done this and made sure everything is level so your potting service does not slant, you can attach the other 4-foot board to the other set of legs.

7. Place the last 4-foot board across the crossbars that are between the legs and nail it in place, connecting the two cross boards to add stability.

8. Lay down one of the 42-inch one-by-four boards across the top of your greenhouse bench. You will cover the entire top of the table with 10 boards that are spaced evenly apart with no more than one inch of space between each board. You have spaces so that the dirt can fall down to the ground, rather than staying on the top of the table. It also allows water to go down to the ground, instead of staying on the wood and rotting it. With these boards, one end will be flush with the top support, while the other end will go half a foot over the edge. This allows you to have an overhang that gives you some extra space.

9. Nail each board into place with two nails on each end. Keep everything aligned so that the top of your potting bench is not crooked.

Potting Bench Plan No. 2

For this, you will need the following along with your wrench, socket, drill, and screwdriver:

- (1) 4 x 8 feet 3/4 plywood

- 15 metal brackets

- 30 bolts 1-1/4 long

- 60 flat washers

1. Lay the plywood on a flat surface, measure 16 inches from one end, and draw a line across the sheet of plywood.

2. Cut along the line so that this piece of plywood will be 16 x 48 inches.

3. Make a diagonal line across this piece of plywood and cut along the line in a triangle. One side should measure 48 inches and the sides should measure 16 inches, making a 90-degree angle.

4. Measure 24 inches from the right along the 48-inch side and make a mark. With a square, draw a line from this mark across the pointed end of the plywood. This line will make a right angle to the 48-inch side of the board. If done right, the line will be four inches long. Make sure you do this to the other triangle piece so they are identical. These are your sidepieces.

5. Cut this line and remove the pointed end piece. Now, you should have a piece that measures 16 x 24 x 4 inches.

6. With the remaining large piece of wood, which measures 80 x 48 inches, measure from the long side to 16 inches, and draw a line from one end to the other. Cut along this line. This will give you a 16 x 80-inch piece and a 32 x 80-inch piece. This will be the bottom and back to the potting bench.

7. Take the 32 x 80-inch piece and stand it on its edge. The top of this piece, as it is now, will be the bottom of the bench. Make five marks on this piece to note where the brackets are going to go. Make sure that the brackets on the end are at least 1-1/2 inches away from the edge.

8. Mount the brackets along these marks you have made. Then, mount the back piece to the bottom piece.

9. Install the sidepieces now.

10. You have now created a legless potting bench. You can install legs if you want but many builders of this type of bench prefer it, because they can get it closer to their pile of potting soil as it is more portable.

11. You can move this anywhere you wish after you have put soil into it to use for potting. You can also put this bench on top of two sawhorses.

Remember, a potting bench can be something as simple as placing a big piece of plywood across two sawhorses. You do not have to get complicated with the potting bench if you do not want to.

Potting Bench Plan No. 3

With this greenhouse plan you will need to be more familiar with woodworking. This plan is for the experienced builder, as you will have to cut pieces of lumber into smaller pieces. For this potting bench you will first construct the sides of your bench, and you will need the following materials:

- (6) 2 x 4s 8 feet long

- (12) 1 x 4s 8 feet long

- (1) 8 x 4 sheet 1/4 plywood

1. The first thing you need to do is prepare the lumber by cutting it into the following pieces:

   a. (4) 6-foot long 2 x 4s

   b. (2) 34-inch long 2 x 4s

   c. (4) 30-inch long 2 x 4s

   d. (2) 14-inch long 2 x 4s

   e. (12) 48-inch long 1 x 4s

2. You will build the sides of your bench first by laying out two 6-foot two-by-fours and one 34-inch two-by-four. Attach a 30-inch board across the bottom of one of the 6-foot boards and the 34-inch board with wood screws. Make sure the corners are square and even.

3. Put the second 6-foot board under the 30-inch board so there is an 8-inch opening between them. Attach the bottom of the 6-foot board.

4. Attach the second 30-inch board across the top of the 34-inch board to where it is the same distance up from the top of the bottom 30-inch board on the back 6-foot board. Attach the 30-inch board to the three boards with wood screws. Attach a third 30-inch board 1-foot below the top, between the two tallest 6-foot boards. Make sure it is the same distance from the top on both boards.

5. Lean one side up on its backboards against the wall, and nail one 48-inch one-by-four shelf piece to the top, middle, and bottom 30-inch boards. Attach the second side of the bench to these boards.

6. Stand up the potting bench and attach a Z brace of some scrap lumber along the back in order to give some brace. Add in bottom shelves by putting the one-by-fours in as bottom and middle shelves. Make sure to leave a 1/4-inch space so that water and dirt can drain through. Attach a top shelf with two more one-by-fours using two boards and sand it all down so the edges are flush.

7. Remove the brace and put a sheet of 1/4-inch plywood as a backer board for the bench. Attach the board to the backboards and sides.

Shelving

Shelves in your greenhouse are a great space saver. Along with having a potting bench, you can use shelves to hold your tools, supplies, and plants. With shelves, you can greatly maximize the overall space within your greenhouse. If space is limited in your greenhouse, having shelving in your greenhouse will help you plant more. You can buy shelves from any gardening shop, or you can custom make your own shelves completely to your needs.

Shelf Plan No. 1

For this you need concrete blocks, lumber, a saw, and a tape measure.

1. Measure out the interior sides of your greenhouse so that you can make a shelving unit that will fit within your greenhouse. If you have a classic A-frame greenhouse, or a hoop greenhouse, you will not be able to build as high as you would if you have a greenhouse with tall, straight sides.

2. Measure out the amount of wood you will need to make three tiers of shelves on the shelving unit and three shelving units in total. You should plan to construct your shelves so they have about 2 to 3 feet in width, and each shelf should have a bit of space between the shelving boards to allow for water and dirt runoff.

3. Your boards should be roughly 8 feet long and 4 to 8 inches wide, and be 2 inches thick. This will keep your shelves from warping when things are sitting on them.

4. With the concrete blocks, lay them on their face so that each block measures 2 feet long and 1-foot high. This will give you the most stability. Ensure you buy enough blocks to support three tiers of shelves on the ends and every 3 to 4 feet in the middle if the boards are too long.

5. Cut the boards to the required length that you want. You can also cut them down the middle to get more shelves.

6. Stack the blocks to the required height for the first shelf. Place the board that you made on top of the concrete blocks, then stack concrete blocks on top of the ends of the shelf. Continue to do so until you have created your two tiers. You should only stack about two blocks at once, so each space between the shelves will be about 2 feet.

Shelf Plan No. 2

To build this shelf unit, you will need the following supplies:

- (8) 2-inch diameter PVC pipes, 10 feet long

- (34) PVC tee connectors, 2-inch

- (4) PVC 90-degree elbow connectors, 2-inch

- PVC adhesive

1. Measure and mark each pipe to make the following:

   a. Eight pieces of PVC that are 4 feet long (front and back of each shelf). Mark these as No. 1.

   b. Ten pieces of PVC that are 21-1/16 inches for the shelf sides. Mark these as No. 2.

   c. Eight pieces of PVC that measure 12-3/4 inches for the top two shelf dividers. Mark these as No. 3.

   d. Eight pieces of PVC that measure 15-1/2 inches for the bottom shelf dividers. Mark these as No. 4.

   e. Sixteen pieces of PVC pipe to measure 1-3/8 inches for the connectors. Mark these as No. 5.

2. Cut the pieces with a hacksaw.

3. The stem of the tee connectors should have the pieces labeled with No. 1 at each end. Ensure that each tee connector mirrors the other, because they are going to be positioned parallel to each other. Continue to do this process by gluing in eight of the No. 2

pieces into the tee connectors. Again, ensure these tee connectors on each pipe are parallel.

4. Put the No. 1 and No. 2 pieces on the floor so they are in the shape of the shelf tops. Place the long No. 1 pieces parallel to each other and the short No. 2 pieces at each end.

5. Turn the No. 1 tubes so that the tee connectors on each end are pointing up and down. Glue in the No. 5 connectors to the bottom of the T cross bar. Continue to do this for all the No. 1 pieces.

6. Connect the No. 1 pieces to the short No. 2 sidepieces and glue them together by gluing the No. 5 connectors to the top of the tee connectors attached to the No. 2 pieces.

7. Assemble the shelves by placing one rectangle piece on the ground and fitting the No. 4 divider pieces into the open side of the tee connector on each corner.

8. Fit a second rectangle over the No. 4 pieces that are sticking up from each corner of the bottom rectangle. Glue these together.

9. Glue the last No. 4 pieces into the tee connectors on each corner. Then glue the rectangle over the ends of the No. 4 pieces that are sticking up.

10. Assemble the shelves using the No. 3 pieces in place of No. 4 pieces.

11. Glue the last four No. 3 pieces onto the open tops of the tee connectors on each corner.

12. Glue an elbow connector to each end of the remaining No. 2 pieces. Twist the elbow connectors in order to ensure they are curving down and fit them with two No. 3 pieces at one end of the shelves. Continue to do this for the other ends of the shelves.

Shelf Plan No. 3

To make this shelving unit, you will need the following:

- Bookshelves

- Plastic sheets

- 2 hinges

- Lathes

1. Purchase a freestanding bookshelf that has no back to it. The open back and front is very important, because it will allow your plants to get the most sunlight.

2. Measure out the height and the width of your shelves so that you can make frames for the plastic sheeting that you will use on the sides of the shelves.

3. Use the lathes that are as wide as the edges of the bookcase to make the frame and then attach the lathes to each other with nails. Hold the frame up to your bookshelf and measure the frame to make sure it will fit within your shelves. Make a frame for the front and the back of your bookshelves.

4. Staple the sheeting to the frame. If you need to, add another lathe to the middle of the frame in order to help support the plastic sheet and keep it from sagging.

5. Attach the hinges to one side of the shelf. You will use this to attach one of the frames so you can create a door that will allow you to water the plants.

6. Attach the second frame to the other side of the shelves using screws or nails.

Raised Gardening Beds

Within your greenhouse, you may not have the garden within the ground, but instead on raised beds. Raised bed gardening is a form of gardening in which the soil is in a container that is 3 to 4 feet wide and can be any type of shape or length. The soil will be raised above the soil around it by about 6 inches to as much as 3 feet. The enclosing frame for the raised gardening beds can be made of many different types of material including concrete, rock, and wood.

Within raised gardening, vegetable plants are often placed in patterns and are closer together than they would be in a typical row garden. This essentially helps to create a microclimate within the gardening bed, because the plants are spaced close enough together that they will barely be touching each other when full grown. Raised gardening beds are great if you practice square foot gardening, where you try to get the most out of the space you have, or companion planting, where you put plants together that help each other grow and ward off predators.

This family has multiple raised vegetable garden beds in their greenhouse. Gardening beds are more productive and easier to maintain than traditional garden plots.

Advantages

There are several advantages to raised bed gardening. These advantages include:

- Raised beds will warm faster in the spring, which allows you to plant and work the soil sooner, thereby allowing your growing season to extend even further.

- The drainage of raised beds will be much better than in typical gardening.

- The soil in raised beds does not get compacted.

- Working the soil and the garden is much easier when you are able to stand and work in the garden than if you are crouched over trying to work within the garden. This helps prevent you from hurting your back, and it also means you can work the garden for longer periods without becoming tired or uncomfortable.

- It is easier to tailor the soil of your raised beds specifically towards the plants that you are planning to grow.

- It is easier to keep predators away from your plants if the plants are raised several feet off the ground.

- Raised gardening beds, once constructed, require much less maintenance than typical garden beds.

Disadvantages

- The timbers may need to be replaced eventually because of the weather. There is pre-treated timber that you can purchase; however, make sure that the chemicals used will not harm your prospective plants, vegetables, or fruits.

- Wall construction costs and time can be a major concern for gardeners looking for an aesthetically pleasing accent piece for their garden. Raised beds may not be the option for you, because this project requires both money and patience.

- Drainage can be a problem with raised beds; during the summer months water is not as plentiful and beds can dry out quicker.

This can lead to stress on your plants, increased chance of disease because of weakened condition, and yield reduction.

- Some individuals find a problem with poor air circulation with some raised beds because plants are closer together. This normally is not a problem but poor air circulation can result in disease ridden species.

- Rotting can also occur in closer quarters if the roots are not allowed to dry properly.

- Moving your raised bed can certainly be done, but it is not recommended, as this could cost you time spent caring for your garden.

- Along the lines of decreased space in raised beds, plant sprawling can also be cause for concern if plants are given too much space. Vines can sometimes grow over beds and into walkways.

Putting Together Your Beds

If you are going to make a site for your raised beds and prepare them within your greenhouse, you need to do the following:

1. Select the area within your greenhouse where you are going to have the raised garden beds. Make sure that, if you are going to be putting in vegetables, herbs, and flowers that like the sun, you select an area of the greenhouse that will get at least eight hours of sun every day. You want your site to be level because this will make it much easier to build on.

2. Make sure that you have easy access to water and that there is enough room for you to work around the bed. It may be difficult to work the raised garden bed if one side is right against the greenhouse wall.

3. Now, you need to figure out the size and shape of your garden bed. If your garden bed is going to be only 6 inches high, then make sure that you can work the bed without having to step on any of the plants. If your garden bed is going to be 3 feet high, then make sure that you can reach all the areas of the garden bed from each side. You do not want to lean over the plants to access the middle of the garden bed, because you could damage plants that way.

4. Prep your site properly by ensuring that you have the right choice for the depth of the bed. If you are growing vegetables or herbs, then 6 inches to 1-foot deep is very important. You want to make sure your plants' roots have enough space to grow, especially for deeper plants like potatoes.

5. Now you need to make the sides of the bed. Use cedar because it is rot resistant, or find a new composite lumber that resists rot. The best sizes are two-by-sixes because they are easy to work with. They are also perfect if you are going to have a bed that is only 6 inches deep. If you are going to have deeper beds, you will need wider boards, or you may have to go for plywood. If you use plywood, reinforce the plywood with two-by-twos or two-by-fours to ensure that the plywood will not break under the pressure of the dirt. Making butt joints on each corner can help strengthen the overall structure as well.

6. Once you have constructed the frame of your raised bed, make sure it is completely level on all sides. If the frame is not level, then the water may run off to one side, robbing plants of vital water and water logging other plants within your gardening bed. You may have to adjust things by removing or adding soil to make things level.

7. Fill up the frame with soil and compost. Mix in some manure, topsoil, and compost. Rake the soil and churn it up to allow oxygen to get in. At this point, you can begin planting seeds and preparing your gardening bed for growing your plants.

Every spring and fall, you will need to top dress new compost and manure on the bed, and churn it up to help keep nutrients in the soil. You will need to mulch the soil to maintain the amount of moisture in it and to keep weeds out. Yes, weeds can get in, but you will not get as many weeds as you would if the plants were growing on the ground rather than on raised beds.

Constructing a Gardening Bed

Because making a gardening bed is universal, we have only one plan that you can customize to make several different types of beds that vary in size, materials, and shape.

1. Plan out the bed that you want to make based on the room in your greenhouse. Put these measurements together into a drawing so you know exactly how large your bed is going to be and how many beds you are going to put within the garden.

2. Choose a material for your raised bed. You can make the raised bed out of lumber, plastic, railroad ties, bricks, rocks, synthetic wood, and anything else that will hold dirt in. By far the easiest material to use is wood, because you can easily shape it to what you want. However, bricks or rocks may look the best and with some time and patience, you can make some nice gardening beds within your greenhouse.

3. Gather the following supplies:

   a. 4 x 4 posts cut to 24 inches in height for your corners

    b.   2 x 12 plywood

    c.   Nails, hammer, compost, soil, seeds, shovel, rake

    d.   Plastic liner for weed barrier

4.   Connect the sides of your bed together to create the shape that you want. You can cut the plywood so you can have three, four, five, or six sides if you want to make the desired shape. Once you have done this, nail and screw into the posts that you have set up for the bed to add strength. If you only screw into the other pieces of plywood, your frame will not be anywhere near as strong as it would if you screwed into the posts.

5.   Cut out a piece of plastic so that it fits into the bottom of your raised bed.

6.   Place the raised bed frame over top of the plastic you have put in the bottom.

7.   Put soil and compost into your raised bed. You should have at least 2/3rd soil and 1/3rd compost. Then mix it up to help ensure that the compost is throughout the raised bed.

8.   Mix in some dry organic fertilizers.

9.   Begin planting!

Give your gardening bed a roof

To help ensure you get the maximum yield out of your raised bed garden within the greenhouse and to eliminate insects, do the following:

1.   Put brackets in the side of your raised bed and slide PVC pipe into each one to create an arch. Each bracket should be about 3 feet from the other bracket.

2. Use spun fiber cloth and clip it to the arches. This helps to conserve moisture while keeping the insects out.

3. When the plants become too tall for the fiber, uncover the portions that need to be uncovered but continue to cover the smaller plants.

4. When things start to become too cold, even in the greenhouse, add some extra heating retention by switching from fabric to plastic.

5. When the temperature within the greenhouse is higher than 90 degrees, take the fabric and plastic off to keep the plants from overheating.

Raised Garden Bed Plans

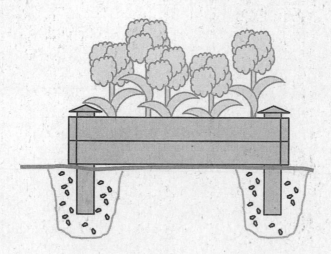

Bill of Material

Item	Qty	Description	Item	Qty	Description
1	4	4x4 x 2'-0"	4	32	1/4" lag screws x 3" LG
2	2	2x6 x 2'-3"	5	4	4x4 post covers
3	2	2x6 x 4'-0"			

Step 1: Install Corner Posts:

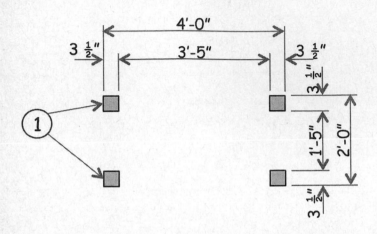

Dig 4 holes 6" diameter by 12" deep. Place corner posts item #1 in hole and backfill.

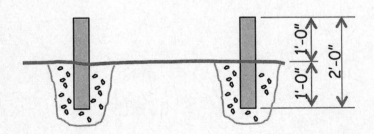

Step 2: Attach Side Rails Corner:

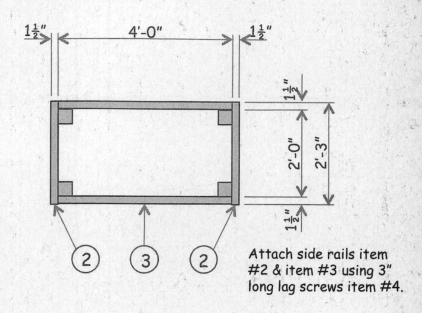

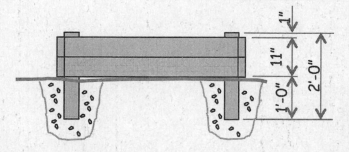

Attach side rails item #2 & item #3 using 3" long lag screws item #4.

Step 3: Finishing:

Install decorative post covers item #5. Fill garden bed 2/3 full of soil and add plants.

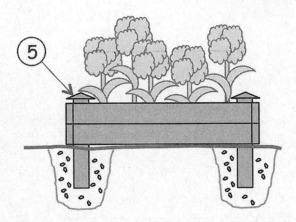

Summary

When you have the aforementioned accessories within your greenhouse, you make your entire gardening experience much more efficient. It is also about personalizing your construction project and making it something that represents your personality when you are inside the greenhouse. Although items such as potting benches and shelving are not necessary fixtures for your greenhouse, they can not only cause you to work more efficiently but also add an aesthetic touch.

Chapter 10

Choosing Covering Materials

> "Gardening requires lots of water — most of it in the form of perspiration."
>
> — Lou Erickson, American writer

I n the previous chapters, you learned all the ways that you can build your beloved greenhouse and briefly discussed the materials you can use to cover your greenhouse. In this chapter, you are going to learn the detailed process of covering your greenhouse and all of the materials necessary for such a venture. The type of covering you choose will depend on a number of factors including how much money you have, your ability to work with the coverings, and the climate in which you live.

Glass

Glass is a very beautiful option for your greenhouse because it is the highest quality, but it is also the highest priced option for your building project. In addition, it is the heaviest material for the greenhouse siding, as well as the most difficult for you to install.

When you install glass, it needs to be double or triple-strength glass in order to withstand the elements, and needs to have a sturdy frame. Another downside to glass is heating, as sunlight that filters through the glass can become more intense and burn plants in some cases.

Now, it may seem like there are several problems with glass for a greenhouse but glass has benefits as well. First, glass can protect plants from harmful UV radiation, and when there is a blizzard or dust storm, your plants should be thoroughly protected. Insects have a very difficult time getting into glass greenhouses, which can greatly reduce your need for pest control. Additionally, as previously mentioned, you can greatly raise the temperature of the greenhouse by using glass, thereby increasing the length of your growing season.

You can get greenhouse glass in a variety of sizes including as large as the entire wall of the greenhouse. Smaller glass panes will have more strength, but you will need to have more framing for all the glass. Larger glass panes are weaker, but you spend less on the framing.

This is an example of an all-glass greenhouse.
Photo courtesy of Rough Brothers, Inc.

When you are thinking of buying glass for your greenhouse, these are some things that you should keep in mind:

• Make sure you check out some offers to ensure you buy something that is durable enough to hold up under harsh weather conditions and is of high quality.

- Always make sure that the greenhouse glass is being attached to a frame of galvanized steel. Galvanized steel is very good at blocking excessive ultraviolet rays.

- Assess the particular plants you intend to grow in your glass greenhouse. Greenhouse glass is very good for seasonal plants, in terms of helping them grow. If you are growing many seasonal plants, then glass may be a good option for you.

Polycarbonate

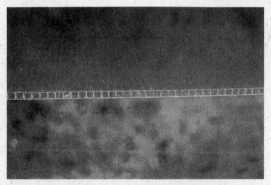

Here is a close-up view of polycarbonate.
Photo courtesy of Rough Brothers, Inc.

Polycarbonate as a way of covering your greenhouse is becoming extremely popular with more and more greenhouse enthusiasts working to install polycarbonate instead of glass for their greenhouse. Glass greenhouses are great, but some of the problems with them, including shattering and condensation, are eliminated with polycarbonate. Polycarbonate is great at insulating a greenhouse and keeping the inside temperature of the greenhouse at a constant temperature. During cooler seasons, this means there is less energy needed to heat the greenhouse when the temperature is cool.

One amazing benefit that comes from polycarbonate is the fact that the corrugated surface of polycarbonate helps to refract light better, thereby giving your plants and vegetation the optimum daylight they need. Compared with glass, polycarbonate is also much lighter, making it easy to construct and work with, and less likely to be damaged during transportation.

Some other advantages with polycarbonate include:

- Polycarbonate helps to create an atmosphere within the greenhouse that aids growing, thanks to constant temperatures, maximum sunlight, and a barrier against wind, rain, and snow.

- Polycarbonate is nearly indestructible and can withstand being hit by hail stones, baseballs, and even rocks.

- Polycarbonate comes in many different sizes and shapes to help you customize the covering for the greenhouse in any way that you need to.

This is an example of a greenhouse made out of polycarbonate.
Photo courtesy of Rough Brothers, Inc.

Acrylic

Acrylic is a clear plastic that resembles glass but in many ways is superior to it. There are two types of acrylic that you can use: extruded and cell cast. Extruded acrylic is made through a process that is less expensive, but unfortunately it is softer than glass and polycarbonate and also scratches easier. Cell cast acrylic is higher quality and is a great choice if you have the money to pay for it. It does not scratch easily and it will stand up to more abuse, which is important if you live in an area where there is severe weather during certain seasons of the year.

Acrylic is heavier than typical plastics but it is only half as heavy as glass, which makes working with it very easy. You can saw it without a problem, making it easier to put onto a greenhouse than glass.

One of the biggest misconceptions about acrylic is that it turns yellow, cracks, and becomes brittle over time. During World War II, this was true in many fighter planes that had bubble-tops made of acrylic, but since then the process of making acrylic has become much more advanced. Nowadays, the acrylic is much better produced and you do not need to worry about it yellowing or breaking suddenly.

As with anything, there are some disadvantages to acrylic and these include the fact that it is more expensive than glass and if there is a direct flame on the acrylic, it will melt and burn. Acrylic will get dirty so you will need to clean it. Detergent and water work well and you will only need to clean about once per year unless you are in an area that has dust storms on a regular basis.

The advantages of using acrylic include:

- It is lightweight and easier to move than glass; therefore, installing it should not be a problem.

- Sunlight filters better through acrylic panels, which creates the perfect conditions for your plants to flourish.

- It is virtually shatter-proof, unlike glass. This allows you a worry free environment to grow healthy plant life.

Fiberglass

Fiberglass is lightweight, it is strong, and it can stand up to hail. This is very important if you live somewhere that has many storms. Hail can break glass, but it will not break fiberglass, which is very important because you do not want to replace it on a regular basis. Typically, fiberglass will last about 15 to 20 years and you will need to put a new coat of resin every decade or so to help keep the fiberglass working properly. If you are going to get fiberglass for your greenhouse, then you should make sure it is clear and made for greenhouses. There is some fiberglass that does not allow much sunlight through, which defeats the purpose of using it in a greenhouse.

Fiberglass is very rigid and will cost you less than glass. It also does not need heavy and extensive structural components, as you would find with a glass greenhouse. However, a disadvantage with fiberglass is that it breaks down under UV light, which is why you need to coat resin over it every decade. Lower grades of fiberglass will need resin about every five years.

Some advantages to using fiberglass include:

- Fiberglass, like glass, will create a greenhouse that is shadow-less, which ensures that all of your plants in the greenhouse get light.

- Fiberglass will also retain more heat than glass does, which will aid your greenhouse in maintaining a constant temperature.

Plastic Film

Plastic is probably the cheapest type of covering you can get but that does not mean it is not going to work for you. Plastic creates a warm and suitable environment for your plants. With plastic film over your greenhouse, you can help protect your greenhouse plants from ultraviolet rays, while still getting natural heat and light within the greenhouse.

There are primarily three types of plastic film that you can get for your greenhouse covering — PVC, polyvinyl chloride, and polyethylene. There are other types of coverings that can be used, but these three are the most common. Polyvinyl chloride is more expensive and it will last as much as five years. If you have PVC plastic over your greenhouse, you need to clean it often, usually every few months, but you can buy it in small sheets that are easier to work with. You can purchase any of these types of plastics in utility grade and commercial grade. Utility grade polyethylene can be bought at a local hardware store, but you will have to replace it on a yearly basis. Commercial grade plastic polyethylene will last longer, usually about two years, because it is treated to filter and block UV rays, which typically break down plastic.

You can get corrugated types of plastic, which have a long life, do not need shading, and are immune to damage from pests like termites, snow, and hail. Roughly 85 percent of the sun's light will get through, which is great for many plants that need plenty of sun but not too much. Certain plants that require a great deal of light may not be able to get as much sunlight as they need.

One disadvantage to plastic covering your greenhouse, beyond the yearly need to replace the plastic, is that ventilation can be a serious problem. Without ventilation, it can get very stuffy in the greenhouse, which will then make it harder for the plants to grow. Poor ventilation will upset the atmospheric balance in the greenhouse as well.

Some advantages to using plastics include:

- One advantage of these plastics is that they are not very expensive. Typically, the price will be about 4 to 16 cents per square foot, depending on the thickness. As a result, it is possible for you to create a 6 x 8 foot free standing greenhouse by just spending $60 on the covering.

- Another great benefit to the plastic covering is that it can resist strong winds and heavy snow, while allowing 98 percent of the sun's light through the plastic.

- Plastic provides a great insulation to your greenhouse, thereby lowering the heating costs you have during the cooler months.

Summary

When you are thinking about the type of exterior you are going to put on your greenhouse, you need to consider how much it is going to cost you. The cost of the exterior is what will usually dictate which you choose to use. If you can afford most types of coverings like glass, or fiberglass, then your next consideration is going to be the type of climate you live in.

If you live in a climate that is quite cool, then you need a material that is going to hold the heat in the greenhouse easily. The less you have to pay for heat, the better off you are going to be with your greenhouse. In addition, if you live in a warm climate, you do not want the heat staying locked in the greenhouse, because you could end up killing your plants from the heat. Therefore, you need a covering that is going to let the light in, but keep heat from becoming a serious problem for your greenhouse.

A good piece of advice is to talk to others who live near you who also have greenhouses. They will be able to tell you what they use, and the success they have had with it. If you do not know anyone with a greenhouse, just do a drive around your area and find individuals who have greenhouses and determine what they used for a covering. Don't be afraid to go and ask them; most people are happy to show off their greenhouse and they will be happy to talk to you about what they used to cover their greenhouse.

Chapter 11

Accessories and Notions

> *"Gardening is civil and social, but it wants the vigor and freedom of the forest and the outlaw."*
>
> — Henry David Thoreau, American writer

One of the biggest mistakes greenhouse rookies make is leaving the greenhouse as it is once they have build the greenhouse and put the benches and pots in. They forget some of the most important parts of the greenhouse, and those are the things that are inside, which some consider frivolous add-ons. However, these can help make your greenhouse growing experience much more successful.

There are a wide variety of items which you can install in your greenhouse, from misting systems and humidistat to lighting meters and even music. Yes, that is right, music can help you and your garden, yet many greenhouse builders do not even think about it.

Sound

The greenhouse looks great. You are ready to garden, but something is missing. The greenhouse is just a big empty glass box unless you can add some atmosphere to it through music. Many people consider the use of music to help plants grow as nothing more than folk knowledge, but studies have been done to determine if music helps plants.

In 1973, Dorothy Retallack published, *The Sound of Music and Plants.* In the book, she detailed the work she did at the Colorado Women's College in Denver using bio-chambers. She would put plants in these chambers and then play particular styles of music, noting the progress and changes in the plants over the course of the experiment.

In the first experiment she found that playing a constant tone for eight hours disturbed the plants to the point where they would die within two weeks. However, using the same tone for only three hours a day caused the second batch of plants to grow extremely healthy.

She followed this experiment with one using radios. In one chamber, she turned the radio to the local rock station; the second chamber had easy listening music. She played the music for three hours a day and found that the chamber with soothing, easy listening music, in it had the healthiest plants. After two weeks, the plants in the easy listening chamber were leaning towards the radio. In contrast, the plants that were subjected to music such as Jimi Hendrix and Led Zeppelin were very weak.

Music for Your Consideration

At this point, you may be wondering what music you should think about using in your greenhouse. First, you need to consider how to get the music to the plants. Depending on how much you want to spend, you can have a complete surround sound system that broadcasts equally to all corners of the greenhouse, or you can have a radio set up that plays music loud enough it can be heard anywhere in the greenhouse.

A good course of action is to determine what music you find relaxing and what music you find irritating when you are trying to calm yourself. Keep in mind the music you find relaxing may be different from the music you listen to on a regular basis. You can enjoy thrash metal, but you probably do not find it relaxing. The types of music to avoid include heavy metal, hard rock, most types of rock, and alternative music. Pop can be okay if it is soft, as is easy-listening music. However, by far the best music to use is classical music. Classical music is not only calming, but plants really seem to enjoy it.

In a study by composer Don Robertson, he played modern classical music to one group of plants, and 16th century classical music to another group of plants. Plants that listened to the older form of classical thrived, most likely because the music from the 16th century composer was far more soothing than the more modern classical music.

According to some research, low frequency sounds can activate enzymes in plants, while promoting DNA replication and increasing the fluidity of the cell membrane in plants. Frequencies between 125 and 250 Hz were found to stimulate a plant's growth the most, while frequencies of 50 Hz, produced plants with less growth. Another study by the Tuscan winery Il Paradiso di Frassina, found that shoot growth and leaf area per vine was higher on vines that had classical music playing near them, versus vines that did not have any music at all. In this study, it was found that the volume does not produce any noticeable changes, but low volume is probably the best as it is more soothing than loud.

Shading

Shade Nets are a fabric with lockstitch construction that defends against fraying, tearing, and sagging. The fabric will not fade, mold, or mildew. You can use it to cover your entire greenhouse or parts of it.
Photo courtesy of Juliana America, LLC.

In hot summer months, shade will be important for your plants. It will give relief from the sun on very hot days, and keep your plants from becoming burned out. Shading is used extensively on greenhouses located in hot climates like India. When you choose shading for your greenhouse, there are several options available to you. One of the most effective options is to have curtains that you can open and close throughout the day to provide your plants with a bit of shading. These are an inexpensive, but they do require you to go out to your greenhouse every few hours to adjust the curtains so your plants do not sit in the shade too long.

Curtain system

Curtain systems will work on one of two systems: a gutter-to-gutter system and a truss-to-truss system. The gutter-to-gutter system involves the curtains being pulled flat across the width of the greenhouse at the height of the gutters. One benefit of this method is that the volume of greenhouse air below the curtain is minimized; therefore, less has to be heated. There is also less installation and maintenance with this method. However, if you have circulation fans or heaters that are above the gutter level, the curtain will block the heat or circulated air from reaching your plants. It is harder to mix and reheat air in this type of curtain system as well.

Truss-to-truss system

The truss-to-truss system involves the curtains moving across the distance between one truss and the next. One advantage of these is that when rolled up or in a bundle, truss-to-truss systems are more compact and minimize the unwanted shade created by the bundle. Installation is easy, and the curtains can be flat at the gutter height, which therefore minimizes the heated area. These types of curtains can follow the slope of the roof as well.

This is an example of a commercial greenhouse that utilizes a truss-to-truss curtain system. It uses curtains along the sides of the greenhouse.

Drive mechanism

If you want to go the extra mile with your curtains, you can get a drive mechanism installed. This is an electric motor and gear box that will extend and retract your curtains for you. A drive mechanism can save you the time and effort of having to go and move each curtain individually, or as a whole with a crank, which can be very tough to do if you have large and heavy curtains. The push-pull drive can be used to move a truss-to-truss system, while the cable-drum and cable-chain drives can be used for both types of curtain systems.

Screening Agents

For the gardener on a budget, a screen agent can be a great alternative to a shading system. Although, it is still a shading option, these agents provide an easy, perfected, and cost effective method for providing you with the shading your greenhouse requires without the hassle.

Temperzon whitewash

Another option is temperzon whitewash, which is a screening agent used on glass and poly houses. You can get this in liquid form and it will dull the light coming into the greenhouse, giving your plants enough light, without too much heat. On days where you want as much sunlight as possible coming into the greenhouse, like rainy days, the whitewash will become transparent when wet.

Decilight®

Decilight® is a calcium-based screening agent that will work on glass and plastic and is one of the best methods of shading you can use to get the optimal screening and maximum heat reduction. In dry conditions, it will produce a screening rate of 80 percent, and a screening rate of 60 percent in wet conditions. Many horticultural greenhouses that have large crops use this method because it is very effective and is perfect for sunny regions. In addition, because it is made up of starches and carbonate, it is not harmful to the environment.

Luxtech®

Luxutech® is a screening agent that is easy to remove if you want to get rid of it, becomes transparent when it rains, and is free of harmful chemicals. In addition, it is long-lasting, which means you will not have to keep replacing it on your greenhouse panels.

If you want to go for the low-tech solution, which is even better than curtains, you can take care about where you are going to put your greenhouse. As mentioned previously, if you build your greenhouse so that it gets partial shade during the day, you will be doing your plants several favors. Your plants get the cooling aspect of the shade, and a break from the sun, which can become very hot in a greenhouse. Look for trees that can give some shading, or even a building that will provide shading from the sun's rays.

By just using a bit of thought and using your creative ingenuity before you put your greenhouse plan into the ground, you can save yourself time and money instead of buying curtains or screen agents.

Misting Systems

One of the drawbacks of a greenhouse is that, because it protects plants from the elements, it keeps out some of the vital necessities your plant needs to survive, such as water. Therefore, you need to provide water for your plants. You can walk around with a hose but this will often lead to wasted water, which means wasted money for you. It also means there is a time commitment for you to go out every day, or at least every few days, to water the plants. Instead, what you could do is install misting systems, which will help your plants grow, save water, and save time for you.

A misting system is also a very effective way to control the humidity within your greenhouse, which is essential to the health of your plants. In a greenhouse, you want to have a humidity level of between 50 and 70 percent, for the best results in growing your plants.

If you do not have a misting system within your greenhouse, it can be very difficult for you to keep the humidity from fluctuating greatly with the temperature changes. If the humidity level drops quickly, the plants can suffer from extensive stress. One of the important things about misting systems is that the water droplets are small enough that they will hang in the air, helping the humidity. However, if the water droplets were to be larger, they would just fall, without helping the humidity level. Depending on how high-tech the system you install is, you can get some systems that will detect the humidity level and either increase or decrease the level of humidity through spraying. One great thing about a misting system is that you can have fertilizer added to the water supply, which will allow the misting system to spray fertilizer on your plants, allowing them to grow better.

The misting system essentially takes advantage of the natural ability of the plants to process water through their leaves and roots. This is probably the biggest advantage of the misting system, along with how it aids in maintaining optimum humidity. There are, however, some disadvantages to misting systems that you should be aware of. Overly damp surfaces in a greenhouse where mist pay pool can create pools of water where bacteria and mold will be formed, which can be harmful to plants. Gray mold, which can cause infection in plants, is one type of mold to be aware of. Mildew can also infect plants and reduce the quality of the vegetation that you get out of your greenhouse garden. Some high end misting system will keep too much water from being emitted, which will then save you from having to deal with mold and mildew.

Mist & Cool® is a do-it-yourself misting and cooling system that can be used with an automatic timer.

Photo courtesy of Juliana America, LLC.

If you are going to get a misting system, you need to consider the type of misting nozzle that you are going to use. The most common type is the brass misting nozzle, which is reliable and will last for many years. If you cannot afford brass, you can go with a ceramic misting nozzle. Some of these nozzles have a stainless steel coating, but most ceramic misting nozzles are made entirely of ceramic. When you get a ceramic misting nozzle, you should find one that creates droplets between 1 and 10 microns in size. The smaller the water droplets the system creates, the more water pressure is needed in your system. Ceramic misting nozzles wear less than other materials, and therefore are a long-lasting product.

One thing to consider, when you have a misting system that has small orifices for the mist, is that the systems can become clogged. Misting nozzles will have a filtration system that will catch the particles, filtering out anything down to 5 microns in size, with another filter system filtering out particles one micron in size. Over time the filters clog and you will have to clean them. However, that being said, typically misting nozzles are low maintenance, and you will not have to check the filters often.

If you do want to clean the misting system, you will need to remove the misting nozzles. When water is left to sit in the nozzle, it can create a deposit that affects the nozzles which can clog the nozzle. If you want to clean it, remove the nozzle from the line and place it in a vinegar solution that is weakened by water. You can buy mist nozzle cleaners if you do not want to use vinegar in the system. Follow these simple steps to clean out your mist nozzle:

1. Leave the nozzles sitting in the vinegar solution overnight, and then take the nozzles out of the solution the next morning.

2. Tap them lightly with a screwdriver to loosen up the nozzle deposits, swirl it around in the solution once more to get any bits of particles still left in the nozzle.

3. Do not use pins or nails to clean out the holes in the misting nozzle because you can damage the nozzle and effectively keep it from working properly.

Grow Lights

It may seem odd to have lights in a greenhouse, because a greenhouse is designed to get the maximum amount of sunlight, but grow lights can be very beneficial in the offseason. During the winter, especially in northern latitudes, there is far less sunlight so you may have to supplement the sunlight with grow lights. In the winter, you may only get three hours of good sunlight for your greenhouse, and the grow light will help you get the extra amount of light that you need. A grow light is essentially an electric light, or a series of lights, that emit an electromagnetic spectrum of light that is beneficial to photosynthesis. These lights emit the same type of light spectrum as you would see from the sun, which allows for indoor growth in a home, and even in a greenhouse.

Incandescent grow lights

Incandescent grow lights use a red-yellow tone and low color temperature, but they are not considered to be true plant-growing lights. They can typically last for 750 hours, are not energy efficient, and give off plenty of heat, which is why they are not energy efficient.

Fluorescent grow lights

Fluorescent grow lights are used for growing vegetables like leaf lettuce, spinach, and herbs. They are often used as a way to grow seeds, and to start spring plantings as well. Unlike incandescent lights, the fluorescent grow light can last for 20,000 hours. You can also get high output fluorescent grow lights that produce double the light of typical fluorescent lights and are useful in areas that are limited vertically in a greenhouse. You can also get compact fluorescent lights that are smaller and used for growing larger plants. They are designed with reflectors that direct light to plants, thereby limiting the wasted light. They can typically last for about 10,000 hours. The last type of fluorescent lights you can get for growing are high-output fluorescent hybrids that do not emit much heat but also have a high intensity discharge technology that allows them to blend light colors and have plenty of coverage over the plants. The electricity costs of these lights are also much less than incandescent lighting.

This greenhouse is using fluorescent lights to grow vegetables.

High-pressure sodium lights

High-pressure sodium lights are used for the second phase of growth for a plant, typically during the reproductive phase when the plant is flowering. The big advantage of these high-pressure sodium lights is that they enhance the fruiting and flowering process within the plants because of the orange-red spectrum that helps plants in their reproductive process. However, sometimes, due to the poor color rendering of these plants, the plants do not appear healthy when they are growing. That being said, these lights often produce high quality herbs and vegetables. These plants have extremely high energy-efficiency and that makes them very popular to use in greenhouses. One thing to be concerned with is the fact that the HPS lamps emit plenty of heat, and create a distinct infrared signature that can attract signatures. In some high pressure sodium lights, you can use a metal halide bulb. This provides the extra bit of colored light for the plants, creating the perfect spectrum blend. These types of lights cost more and they have a shorter lifespan.

Convertible two-way lamps

Convertible two-way lamps use a metal halide bulb or a high pressure sodium bulb. Unlike the above example, you cannot have both of these lights in the fixture at once. These are very easy to change out when you need to; all you need to do is switch the bulb off, set the right setting, and put the new bulb in.

LED grow lights

LED grow lamps provide cheap, bright, and extremely long-lasting light for your plants. Because these do not consume much power, they are highly beneficial for use in greenhouses during the winter when they may need to be on for several hours. One big benefit of these lights is the lack of heat they emit, which means you have to water less because the soil around the plant will not become dried out as quickly. Of course, make sure you do not overwater the plants. New LED grow lamps use grade 6 watt LEDs that create the same type of power seen in HID lamps. You can typically do the following when replacing your other grow lamps with an LED lamp:

- A 90 watt LED light can replace a 400 watt Metal Halide or HID grow light.

- A 120 watt LED can replace a 600 to 800 Metal Halide grow light.

- A 300 watt LED light can replace a 1,330 watt Metal Halide light.

- A 600 watt LED light can replace a 1,600 watt HID light

When you are using grow lights for your plants to supplement the lack of sunlight you may receive in the winter or fall, keep the following suggestions in mind:

- It is important to remember that the bigger the plant gets, the more light it is going to need. If it does not get enough light, the plant will not grow.

- Vegetables grow best in full sunlight, which means you need to mimic the light of the sun as closely as possible. Fluorescent lamps and metal halide lamps work the best.

- Foliage plants grow best in full shade, and therefore do not need to have grow lights over them.

- Plants need to have dark and light periods, and you should have a timer that switches off the lamps for a time. The amount of photo dark period for the plants will depend on the plant. Some plants grow better when the days are long (more light) and the nights are short (less light), while other plants prefer to have the opposite amount of light.

You should know about light density, measured in lux, when you are using grow lights in your garden. One lux equals one lumen of light falling on one square meter of planting area. If an area is brightly lit, it will be illuminated for about 400 lux.

Temperature Gauges

An often forgotten piece of greenhouse building is the temperature gauge. The temperature gauge is very important, because it will keep your plants from becoming victims of heat damage. Too much heat can cause your plants to die, and a temperature gauge will alert you when you may be running into a danger zone for heat. Sometimes called a temperature monitor, this gauge can be very simple or very high-tech. The simplest type of temperature monitor will emit a beeping noise within the greenhouse and hopefully you will hear it as you are walking by. Higher end temperature monitors will send you a text to your phone when the heat gets too high, or even call you or a backup number you have set up if you are out of town. Going even further, some temperature monitors will alert you about power failures, humidity problems, and even intruders into your greenhouse. You can also get a temperature monitor that will allow you to call the monitor from your phone and get an update on what the temperature in the greenhouse is, no matter where you are in the world.

This may seem like a frivolous expense, but when you lose your plants because the heat becomes too much and no one has opened a vent, you will know just how valuable these are. If you cannot afford a temperature monitor in your greenhouse, then you should at least look at getting a thermostat. You can get a thermostat for only a few dollars, and all you have to do is make the commitment to go check the thermostat every day to make sure that your greenhouse is not suffering from excessive heat. At the very least, you should have a thermometer in your greenhouse.

Humidistat

A humidistat is an instrument that will measure and control the humidity within your greenhouse to ensure the greenhouse does not have drastic changes in humidity, which can be disastrous for your plants. Many greenhouse enthusiasts will combine a humidistat with a thermostat and an air conditioner in order to have complete control of the greenhouse mini ecosystem. A humidistat will be used in conjunction with fans, humidifiers, and dehumidifier. The way that the humidistat will control the humidity within a greenhouse can be done through several methods, depending on the system that you have implemented within the greenhouse.

- Changing the speed, or switching on and off a fan or vent blower in order to bring in more air, or removing air to adjust the humidity.

- Activating or deactivating the humidifier.

- Changing the openings of dampers to change the humidity within the greenhouse.

- Switching a dehumidifier on and off, depending on whether or not the situation requires it.

When the humidistat works, it does so in the following manner:

1. You set the humidity to a comfortable level for your plants that will maximize the growing potential for them.

2. When the humidity level goes below what you set on the humidistat, the humidifier will be switched on.

3. As the humidity level changes within the greenhouse, and returns to the normal level you have set, the humidifier will switch off.

4. When the humidity level goes above what you have set for the greenhouse, the humidistat will kick in and start the dehumidifier.

5. The dehumidifier will lower the humidity level within the greenhouse and, when it reaches the right level, switch off.

Lighting Meter

Lighting meters are an important part of your greenhouse plans, because the right amount of light can ensure that your plants prosper and thrive. With the lighting meter, you will know how much light your plants are receiving, and whether or not you have to adjust your shading or grow lamps. A light meter will measure the amount of light in an area. Often, a light meter will include a computer, which allows you to program alerts if the light changes to something that is not beneficial to your plants. Understanding the amount of light your plants need to survive is important when you are using a lighting meter. One important thing to remember when you are dealing with changing the level of light for your plant is to do it gradually, because sudden changes in light can shock your plants. Plants in general will utilize morning light the best, because the plant metabolism is most active in the morning. Of course, if the region in which you live often has fog in the morning, then the plants are going to get the best light in the afternoon. You should watch your greenhouse plants very carefully to avoid sunburn on your plants. Also, species grown in light that is too low for them will not flower or bloom.

Light meters use the measurement of foot-candles (fc) and you can probably expect the following values depending on how much light is coming in.

- 5,000 fc: Full sunlight during midday will register this high on your light meter. In some lower latitudes, you may find that the level may increase to 10,000 fc. Sunlight going through fiberglass will reach 5,000 to 7,000 fc.

- 4,000 fc: Bright light, about half what the midday sun would read on your light meter. This is often the most efficient level of light for your plants.

- 1,800 fc: Light or dappled sunlight will read this much on your light meter.

- 1,000 fc: Reduced sunlight, which typically does not produce a shadow, will register this much on your light meter.

- 500 fc: When your plants are in deep shade, the light meter will register this much in the greenhouse.

Artificial light will register low on the light meter, but that does not mean your plants are not getting the proper amount of light to allow them to grow. One thing to remember with artificial light is that the light is constant. Therefore, though the light level is low, it is spread out over several more hours.

Automatic Vent Opener

The perfect greenhouse is one that manages itself. When the greenhouse is automatically controlled, you are able to watch your plants grow without worrying about the humidity affecting your plants. One more thing to consider with your greenhouse is the vents. Purchasing automatic vent openers can help maintain the temperature and humidity level within your greenhouse.

A roof vent will provide heat with an escape route out of the greenhouse, while at the same time providing a way for cool air to get into the house. Automatic roof vents have the additional bonus of allowing the intake of cool air to be a gentle flow, rather than a sudden blast of air that causes the temperature to suddenly go down. You do not have to use just roof vents though, as side venting will assist in the cooling of the greenhouse during the warm season.

An automatic vent opener works without electricity because it is self-contained. On a sunny day, heat will build up in the cylinder mechanism of the vent opener. This will cause the vent to slowly open, rather than opening suddenly. When the weather cools, the vent will slowly cool as the air around the cylinder decreases in temperature. It is important that when you do have a greenhouse automatic vent opening system, you use it in conjunction with greenhouse exhaust fans. A greenhouse exhaust fan can help drive air out of the vents, keeping the temperature within the greenhouse at the optimum temperature. If you are growing orchids, chrysanthemums, or tropical plants, then you may need an exhaust fan to maintain the controlled environment within the greenhouse. With an exhaust fan, you should have it constantly distributing air throughout the greenhouse, turning slowly to maintain the air circulation. Condensation within the greenhouse can actually be reduced greatly when you have air circulation.

Ground Cover

A ground cover is a black and woven polypropylene fabric that will control weed growth, while at the same time retaining moisture within the soil. Because it is woven, water and nutrients can still get down into the soil, preventing puddles of water from forming. Although you can use black fabric to help keep weeds away, you may want to think about using white on black, because this will reflect photosynthetic active radiation back to the leaves of the crop, which then will maximize the total energy that the plant produces, helping the plant grow even more. Ground cover is very important in a greenhouse, because, though you may not think you will get weeds in your greenhouse, there is a chance that they can find their way in. As a result, you will want to keep weeds from growing all together, which is why ground cover is very important. Although chemicals are a useful alternative, the fumes could harm the potential growing environment. You can use chemicals but most gardeners are moving away from that in order to "go green."

When you use the items discussed in this chapter, you can ensure you will get the optimum growing environment out of your entire greenhouse, thereby ensuring you have the most vegetables and herbs possible.

To recap, these are the items you should consider having in your greenhouse:

- **Shading:** The sun can be harmful for your greenhouse and shading can help moderate the temperature within the greenhouse. Whether you use coating that goes on the glass, a curtain system, or you just build your greenhouse so it is partly shaded by a tree, shading will help your greenhouse and the plants inside.

- **Misting Systems:** Humidity is vitally important within a greenhouse and therefore you need to think about using misting systems. Misting systems will keep your plants hydrated, regulate the humidity, and provide you with an easy method to transfer fertilizer to your plants, helping them grow even more.

- **Grow Lights:** If you are in northern latitude, or in an area like the Pacific Northwest where there are plenty of cloudy days and rain, grow lights can be a saving grace in a greenhouse. The grow lights will help your garden grow when there is not much light. In the winter this is especially important. When you choose LED lights, you save money and that will help make growing items in your greenhouse all the better for you.

- **Temperature Gauge:** If you are not near your greenhouse, how do you know how warm it is in there? The temperature gauge can be as simple as a thermometer or as high-tech as a temperature gauge that calls you when the temperature gets too high. If it is too hot in your greenhouse, your plants will die and you do not want that to happen. That is why something like a temperature gauge is so important for you and your greenhouse.

- **Humidistat:** This might be something you have not heard of before, but it is important for your greenhouse. With a humidistat, your greenhouse is monitored for its humidity level and you can ensure that if the humidity gets too high, or too low, the humidifier or the dehumidifier is going to kick in and therefore maintain the optimal humidity for your plants.

- **Light Meters:** Knowing the right amount of light for optimal growth is all part of being a greenhouse enthusiast. A light meter is cheap, and it can give you an indication if your plants are receiving too much or too little light.

- **Automatic Vent Opener:** One of the most low-tech and easiest things to install in your greenhouse is the automatic vent opener. Using nothing but power and heat, with no electricity, you can rest assured that when things get too hot in your greenhouse, the vent is going to open to help equalize the temperature and humidity. When you add in an exhaust fan, you can ensure there is a constant circulation of air within the greenhouse.

- **Ground Cover:** Ground cover is something that you may not think of when you are building a greenhouse. Greenhouses keep the weeds out, not bring them in. However, you can bring them in with your clothes, the dirt, and the compost that you bring in for your greenhouse plants. Weeds take away vital nutrients and minerals from your greenhouse plants and vegetables, so you want to use woven black and white ground cover to keep weeds from growing, while also making sure that your plants grow. The white reflects vital light back to your plants, giving them an extra boost of energy as well.

The important thing to remember is that much of the technology discussed within this chapter is completely optional. In no way is it completely necessary to purchase the items previously mentioned in order to have the most successful greenhouse. Although there is a level of convenience in automatic systems, manual options are equally sufficient for your greenhouse. However, utilizing the items covered in this chapter can ensure that you get the most out of your growing experience.

Chapter 12

Greenhouse Maintenance

N ow that you know what goes into the construction of your greenhouse and some of the accessory you can put inside of it, it is time to discuss greenhouse maintenance. In order to maintain your newly built space, there are a few monthly, even weekly, chores that must be done. It is especially important to incorporate these chores if you have a misting system, heating system, or any other technological accessory, as these are an integral part of the health of your plant life.

Cleaning and Disinfecting Your Greenhouse

If you have reoccurring problems with diseases such as root rot and gnats and fruit fly infestations, your potting area and benches may be in need of some cleaning. Attention to greenhouse sanitation and disinfection is a step all gardeners can take full advantage of before the growing season. You should begin by thoroughly cleaning the floor of any soil, organic debris, and weeds. To counteract these problems, you can install weed mats, if you do not have a concrete foundation. Weed mats act as a barrier against weeds and help to manage the growth of algae. You should avoid placing anything on top of the weed mat, as soil and moisture could become trapped, causing mold and disease to spread to your plants. Potting trays should always be treated with disinfectant before placing new soil inside them. Work benches and tables should also be made of a non-porous surface such as a laminate; otherwise, moisture can become your worst nightmare. Be sure that you do not over water your plants to avoid puddles and — ultimately — unwanted problems.

Checking Windows and Doors

It is not enough to just keep the inside of your greenhouse clean; it is also very important to make sure the doors and windows you have installed are secured properly. You should do monthly checks of your windows and doors to make sure they are still in working order and that there are no cracks or leaks where air can escape. Make sure that each door and window opens in the correct manner and stays open when necessary. It would be a shame to come back to a greenhouse that was either too cool or too hot because of a malfunctioning entrance.

It is also very important to make sure that the seals on your windows and doors are keeping out weather that could be potentially harmful to your plant life. Making sure your windows and doors are sealed properly may be the difference between healthy plants and diseased ones. Additionally, checking for this goes along with looking carefully at hinges and locks.

Hinges and locks

When checking door and window hinges, you should keep in mind the kind of material it is made from. It is not surprising that, over time, hinges will rust due to the elements. It is not necessary to check these weekly, or even monthly, but be aware that rusting may occur after a year or two if you live in a region that receives an ample amount of rain during the summer months. Unless you are using PVC hinges, annual checks should be performed. If, at some point, you are in need of a door hinge replacement in your greenhouse, follow these simple steps. For this traditional door hinge project you will need:

- Screw driver

- Electric drill

- Slip joint

- Straight claw hammer

- Wood glue

- Golf tees (optional)

1. Remove hinge pins with your electric screw driver.

2. Remove the pins once they come loose. You may need to unscrew by hand, once the pins are loose enough.

3. Lift the door free, but be careful because it may come loose without any help. You may need help with this step, as some doors may be too heavy to lift alone.

4. Remove the two halves of the set of hinges by removing the mounted screws. Do not discard these; you will need them as reference when you purchase replacements at your local hardware store.

5. Once you get your newly purchased hinges home, you will need to split them. This can be done by holding one end of the hinge with one hand then using pliers on the pins.

6. Use the mounting screws to put each hinge half in its place. Repeat this on each set of hinges.

7. With the pins close by, set the hinges into place and slide the door back into aliment. Slide the pins in and tap them in with a hammer to make sure they are completely set.

Checking Ventilation Systems

As discussed in Chapter 3, fans and vents are the typical ventilation systems for greenhouses, and sometimes they need constant care, because dust, mold, and mildew can become a major problem. If you have a fan installed in your greenhouse, it is recommended that you dust weekly to prevent the building up of dirt and dead skin.

Fans and vents

Without dusting, your plants could become susceptible to diseases and die as a result. Dusting can been a simple fix that could save you time and money. The accumulation of only ounces of dust can create enough of an imbalance to reduce operating efficiency of your fan blades. If you have vents in your greenhouse, cleaning them may not be as easy as simply dusting weekly. Whenever cleaning the fans, lubricate the fan bearings, motor, and shutters. Any parts that do not move freely should be replaced immediately. When looking at your fans, be sure to check your fan wheel for proper rotation. Fan rotation can become reversed if a fan is installed improperly, repaired, or when the polarity of wiring circuits is alternated. Proper rotation direction is generally marked on the fan housing, and if you are not sure of this, call your local home repair store. Home Depot and Lowes associates can give you any further advice about this.

Checking Water Tanks and Irrigation Systems

An irrigation system is a very important part of your greenhouse, as manual watering can become tedious, but it can also be a hefty job when it comes to cleaning as well. Depending on what kind of irrigation system you have, there are different measures you will take when cleaning them out. If you have a fan, you can use a simple cleaning agent to get rid of mildew. With a drip irrigation system, particles that are trapped can be flushed out of the filer by opening the faucet. Many of these systems will need to be flushed out daily, especially if you are using fertilizers to stimulate plant growth.

Disinfecting water and killing algae

Clean water is essential to the abundance of plant life in your greenhouse; keeping your water as clean as possible is of the utmost importance. The use of chlorine dioxide will help disinfect your water, but — unless you have a re-circulating system — you should always have fresh water in your irrigation system. Chlorine dioxide can be purchased at Amazon (**www.amazon.com**), REI (**www.rei.com**), and Campmor (**www.campmor.com**).

Copper ionization is another disinfectant used to kill algae and completely purify water. This disinfectant is used by many chlorine-free pool owners to kill bacteria that could be potential harmful. This chemical is completely safe for humans, but be careful getting it on or near your plants. Though it has not been proven as harmful to plants, it is used simply to purify your water supply. Do not use it on your plants as a disinfectant.

Chapter 13

Additional Greenhouse Plans

Greenhouse Plan
Window Mount
12x36x48

 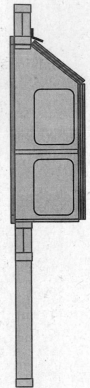

Bill of Material

Item	Qty	Description	Item	Qty	Description
1	1	2x4 x 4'-3"	13	1	1/4" acrylic x 2'-10" x 4'-0"
2	2	2x6 x 4'-3"	14	2	3/8" plywood x 1'-4 1/2" x 4'-0"
3	2	2x4 x 4'-0"	15	1	1/4" acrylic x 1'-4 1/2" x 4'-0"
4	2	2x6 x 4'-0"	16	1	1" plywood x 1'-0" x 3'-10"
5	2	2x4 x 7"-2"	17	1 box	#8 wood screws x 1 1/2" LG
6	4	2x4 x 2'-7 1/2"	18	1 box	#8 wood screws x 3/4" LG
7	1 box	3" 10d nails	19	1 tube	silicone caulking
8	1	1" plywood x 1'-4" x 4'-0"	20	2	10 trim cut to suit
9	4	3/8" plywood x 1'-4" x 3'-8"	21	1	10 ft flashing cut to suit
10	2	1/4" acrylic x 1'-4" x 3'-8"	22	1 box	1 1/2" finishing nails
11	1	1" plywood x 4" x 4'-0"	23	1 qt	primer
12	2	3/8" plywood x 2'-10" x 4'-0"	24	1 qt	paint

Step 1: Frame Wall Opening:

Remove existing wall studs to create a 4'-3" wide opening.
Frame as shown using 3" long, 10d nails, item #7.

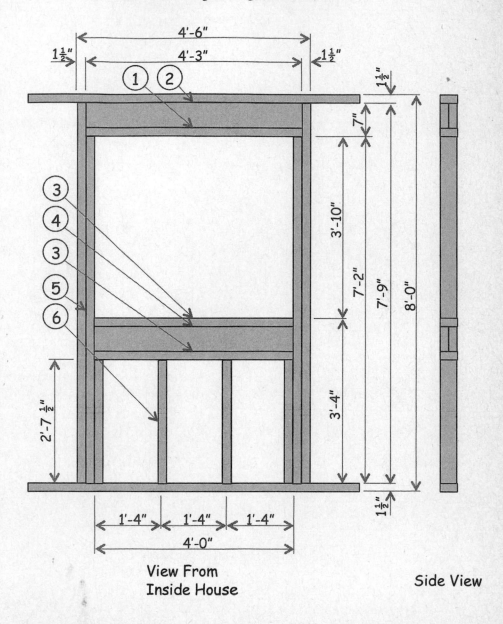

View From
Inside House

Side View

Step 2: Cutout Components:

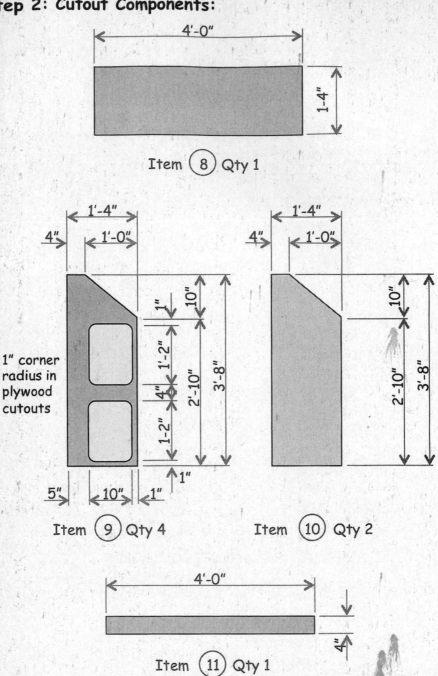

Item (8) Qty 1

1" corner radius in plywood cutouts

Item (9) Qty 4

Item (10) Qty 2

Item (11) Qty 1

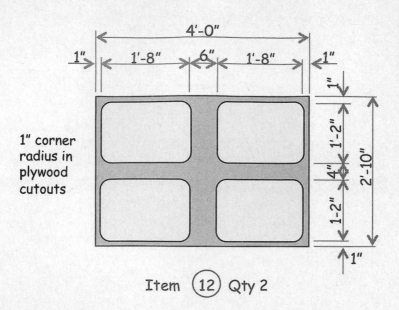

1" corner radius in plywood cutouts

Item ⑫ Qty 2

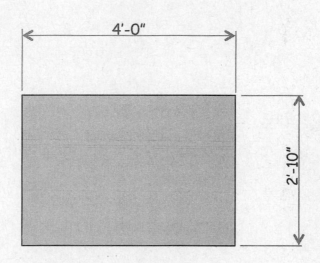

Item ⑬ Qty 1

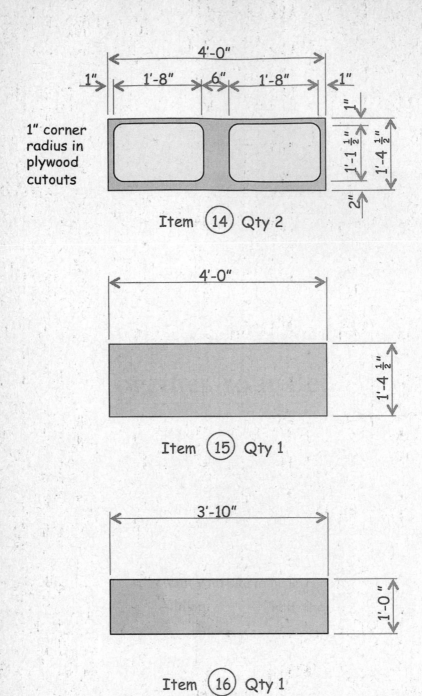

1" corner radius in plywood cutouts

4'-0"

1" 1'-8" 6" 1'-8" 1"

1"

1'-1 1/2"

1'-4 1/2"

2"

Item ⑭ Qty 2

4'-0"

1'-4 1/2"

Item ⑮ Qty 1

3'-10"

1'-0"

Item ⑯ Qty 1

Step 3: Assemble Plywood Panels to Clear Acrylic Glass:

Align item #12 & #13 and attach with ¾" wood screws item #18. Seal all openings with silicone caulking item #19.

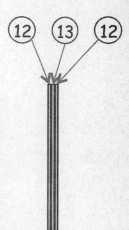

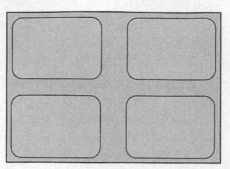

Align item #14 & #15 and attach with ¾" wood screws item #18. Seal all openings with silicone caulking item #19.

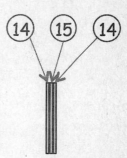

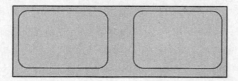

Align item #9 & #10 and attach with ¾" wood screws item #18. Seal all openings with silicone caulking item #19.

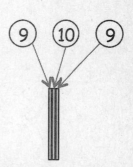

Step 4: Assemble Greenhouse:

Using #8 wood screws × 1 ½" long, item #17 assemble greenhouse components in the following order: item #8 bottom board, items #9&10 side wall assembly, items #12 & 13 front wall, items #14 & 15 roof, item #16 shelve, item #11 top plate.

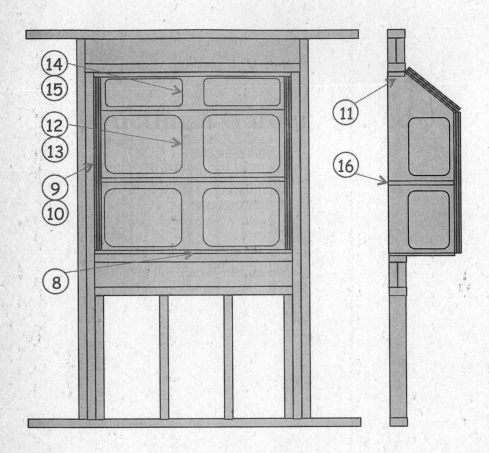

View From
Inside House

Side View

Step 5: Install Trim and Flashing:

Install trim item #20 using finishing nails item #22.
Install flashing item #21 to seal top and sides of
greenhouse. Prime and paint to suit.

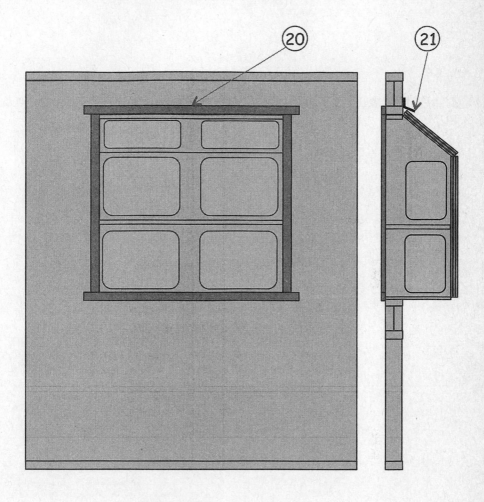

View From
Inside House

Side View

Greenhouse Plan
6x12
Lean-To

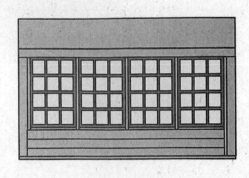

Bill of Material

Item	Qty	Description	Item	Qty	Description
1	6	2x6x 12'-2 1/2"	12	4	2x6 x 5'-8 1/2"
2	8	2x4 x 2'-8"	13	7	32x48 prehung double hung windows
3	2	2x4 x 7'-5"	14	1	2'-6" x 6'-8" prehung door
4	1	2x4 x 6'-10 1/2"	15	8	siding x 10' long cut to suit
5	10	2x4 x 5'-8 1/2"	16	1 box	6d x 2" lg nails
6	10	2x6 x 7'-0"	17	3	1/4" x 4'-0" x 8'-0" acrylic sheet cut to suit
7	4	2x6 x 2'-11"	18	1 box	#8 wood screws x 1" lg
8	2	2x6 x 5'-8 1/2"	19	1 box	10d x 3" lg nails
9	7	2x4 x 1'-7"	20	1 gallon	primer
10	2	2x6 pressure treat x 5'-10"	21	1 gallon	paint
11	1	2x6 pressure treated x 12'-2 1/2"	22		

Step 1: Assemble Rough Framing:

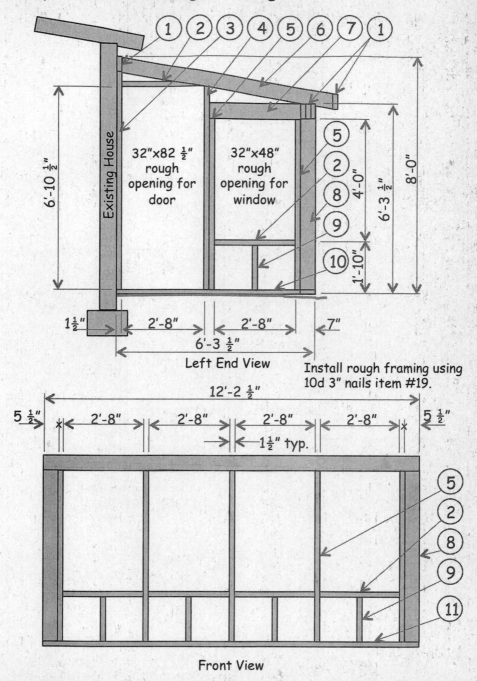

32"x82 ½" rough opening for door

32"x48" rough opening for window

6'-10 ½"

Existing House

8'-0"

6'-3 ½"

4'-0"

1'-10"

1½"   2'-8"   2'-8"   7"

6'-3 ½"

Left End View

Install rough framing using 10d 3" nails item #19.

12'-2 ½"

5 ½"   2'-8"   2'-8"   2'-8"   2'-8"   5 ½"

1½" typ.

Front View

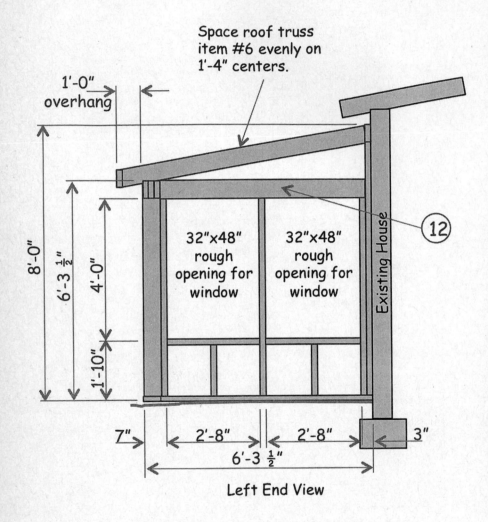

Space roof truss item #6 evenly on 1'-4" centers.

1'-0" overhang

8'-0"

6'-3 ½"

4'-0"

1'-10"

32"x48" rough opening for window

32"x48" rough opening for window

Existing House

12

7"

2'-8"

2'-8"

3"

6'-3 ½"

Left End View

Step 2: Install Windows & Doors:

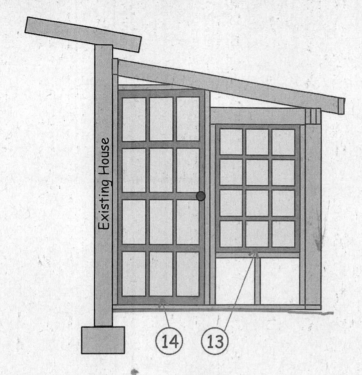

Existing House

⑭ ⑬

Left End View

Install windows and door
per manufacturers
recommended instructions.

⑬

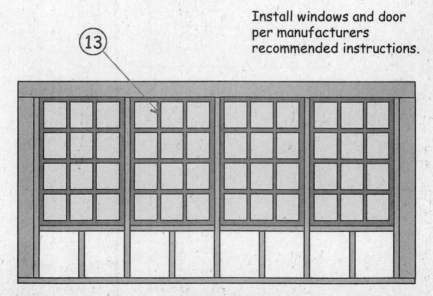

Front View

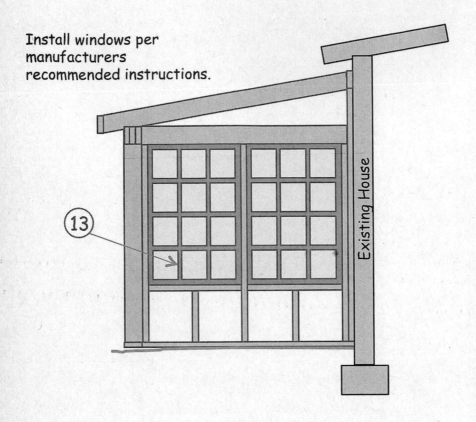

Install windows per manufacturers recommended instructions.

Existing House

Left End View

Step 3: Siding:

Attach acrylic roof panels
item #17 using #8 wood
screws x 1" long item #18.

Install siding item
#15 using 6d nails x
2" lg. Cut siding to
suit as required.

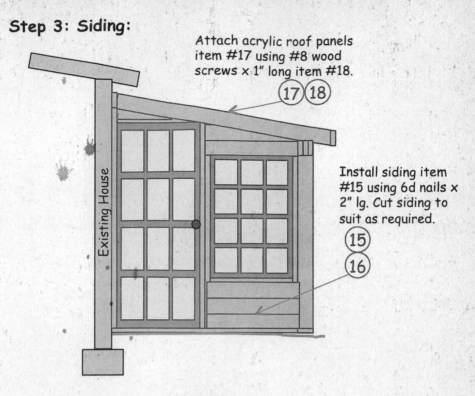

Left End View

Front View

Prime and paint to suit.

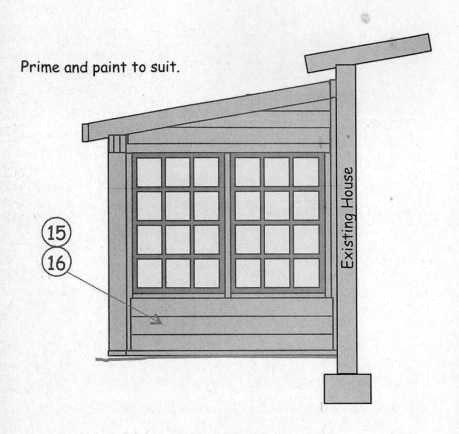

Left End View

Greenhouse Plan
9x12
Freestanding Wood Frame

 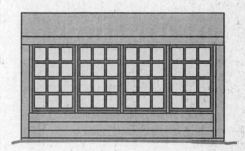

Bill of Material

Item	Qty	Description	Item	Qty	Description
1	11	2x6x 12'-2 1/2"	11	1	2x6 x 8'-9
2	14	2x4 x 2'-8"	12	13	32x48 prehung double hung windows
3	2	2x4 x 7'-0"	13	1	2'-6" x 6'-8" prehung door
4	20	2x4 x 5'-8 1/2"	14	14	siding x 10' long cut to suit
5	20	2x6 x 6'-0"	15	1 box	6d x 2" lg nails
6	2	2x6 x 2'-11"	16	6	1/4" x 4'-0" x 8'-0" acrylic sheet cut to suit
7	4	2x6 x 5'-8 1/2"	17	1 box	#8 wood screws x 1" lg
8	13	2x4 x 1'-7"	18	1 box	10d x 3" lg nails
9	2	2x6 pressure treat x 8'-9"	19	1 gallon	primer
10	2	2x6 pressure treated x 12'-2 1/2"	20	1 gallon	paint

Step 1: Assemble Rough Framing:

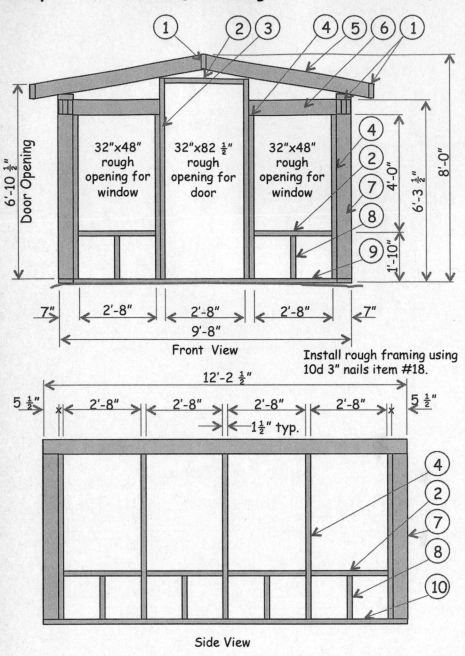

Front View

Side View

Install rough framing using 10d 3" nails item #18.

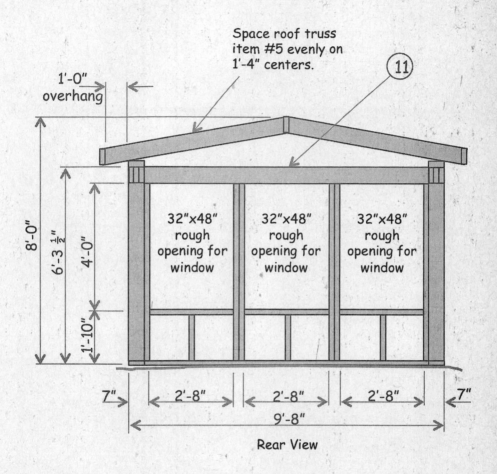

Rear View

Step 2: Install Windows & Door:

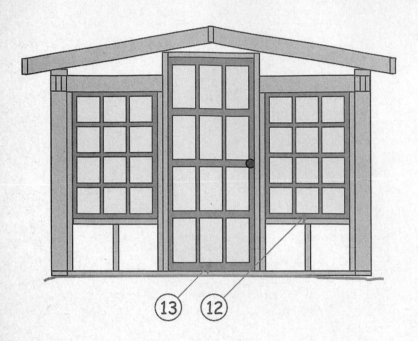

Front View

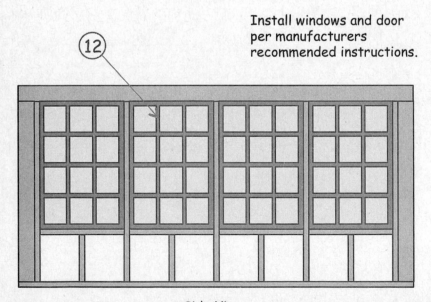

Install windows and door per manufacturers recommended instructions.

Side View

Install windows per
manufacturers
recommended instructions.

Rear View

Step 3: Install Siding:

Attach acrylic roof panels item #16 using #8 wood screws x 1" long item #17.

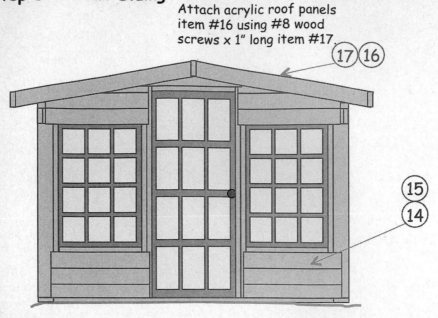

Front View

Install siding item #14 using 6d nails x 2" lg. Cut siding to suit as required.

Side View

Prime and paint to suit.

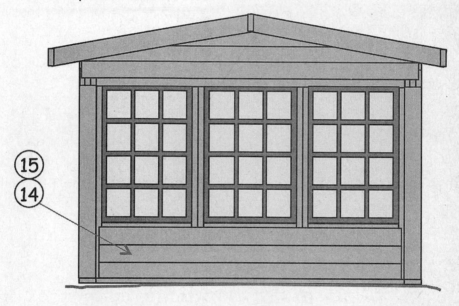

Rear View

Greenhouse Plan
9x12
PVC Hoop Design with
Wood Base Frame

Front View

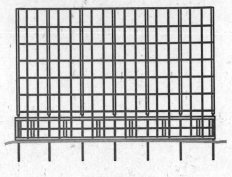

Side View

Bill of Material

Item	Qty	Description	Item	Qty	Description
1	1	2x4 x 2'-8"	10	1 box	10d x 3" LG nails
2	2	2x4 x 6'-9"	11	14	1/2" rebar x 3'-0" LG
3	2	2x4 pressure treated x 8'-6"	12	7	3/4" dia. PVC pipe cut to suit
4	2	2x4 x 2'-11"	13	1 roll	chicken wire
5	25	2x4 x 1'-0"	14	1 box	8" electrical wire ties
6	2	2x4 pressure treated x 12'-0"	15	1 roll	UV stabilized polyethylene sheet
7	2	2x4 x 12'-0"	16	1 gallon	primer
8	1	2x4 x 8'-6"	17	1 gallon	paint
9	2	2x4 pressure treated x 8'-6"			

Step 1: Assemble Base Framing:

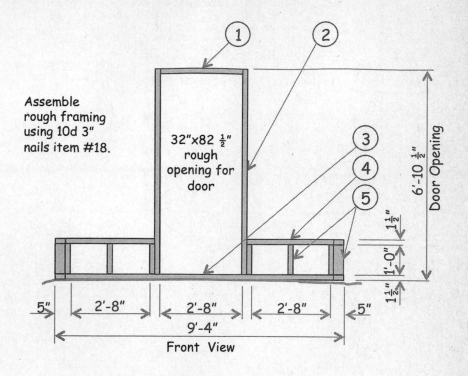

Assemble
rough framing
using 10d 3"
nails item #18.

32"x82 ½"
rough
opening for
door

Front View

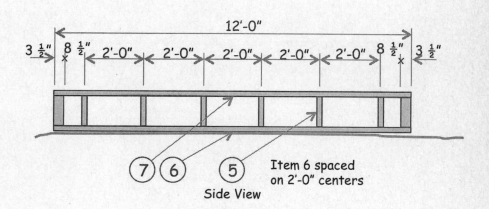

Item 6 spaced
on 2'-0" centers

Side View

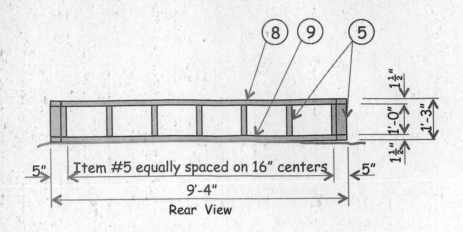

5"   Item #5 equally spaced on 16" centers   5"

9'-4"

Rear View

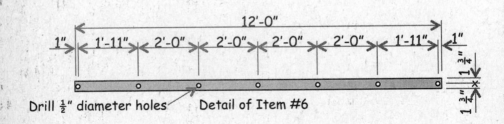

Drill ½" diameter holes   Detail of Item #6

Step 2: Install Rebar:

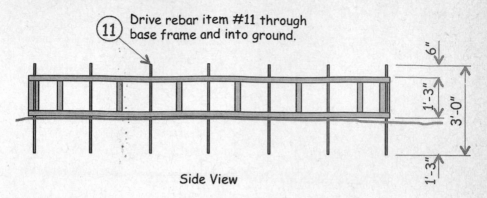

Drive rebar item #11 through base frame and into ground.

Side View

Step 3: Install PVC Pipe:

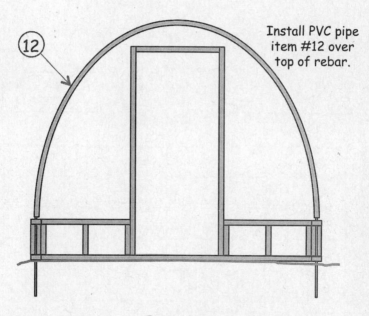

Install PVC pipe item #12 over top of rebar.

Front View

Step 4: Install Chicken Wire & Plastic:

Attach chicken wire item #13 to PVC pipe item #12 using electrical wire ties item #14

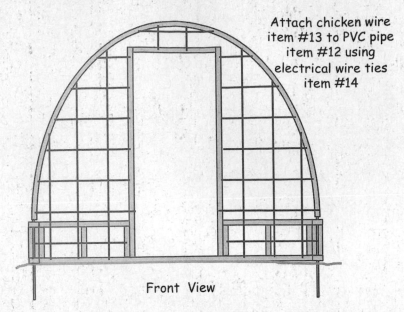

Front View

Cover chicken wire with UV stabilized polyethylene sheet item #15. Staple sheet to wood structure. Paint or stain wood prior to installing sheet.

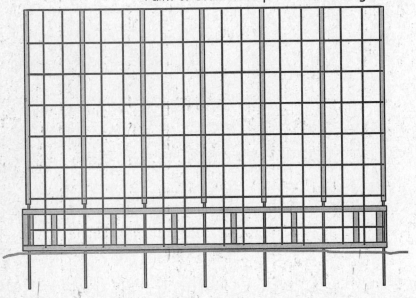

Side View

Step 5: Install Door:

Install pre-hung
door item #15

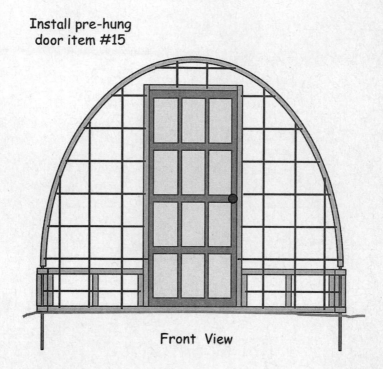

Front View

Greenhouse Plan
12x12
A-Frame

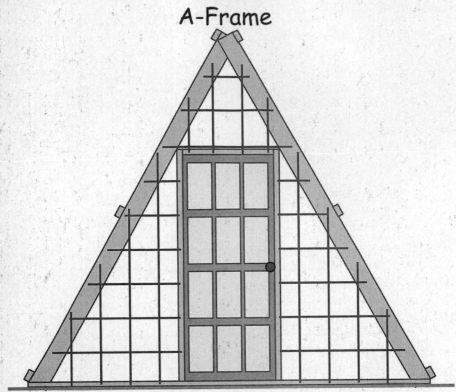

Bill of Material

Item	Qty	Description	Item	Qty	Description
1	1	2x4 x 2'-8"	6	1 box	10d x 3" LG nails
2	2	2x4 x 6'-9"	7	1 roll	chicken wire
3	14	2x6 x 12'-0"	8	1 box	heavy duty staples
4	6	2x4 x 12'-11"	9	1 roll	UV stabilized polyethylene
5	4	2x6 x 12'-0" pressure treated	10	1	2'6" x 6'-8" prehung door

Step 1: Assemble Rough Framing:

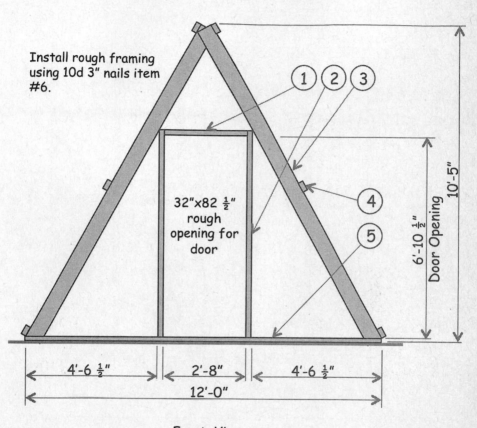

Install rough framing using 10d 3" nails item #6.

32"x82 ½" rough opening for door

① ② ③

④

⑤

10'-5"

6'-10 ½"
Door Opening

4'-6 ½"   2'-8"   4'-6 ½"

12'-0"

Front View

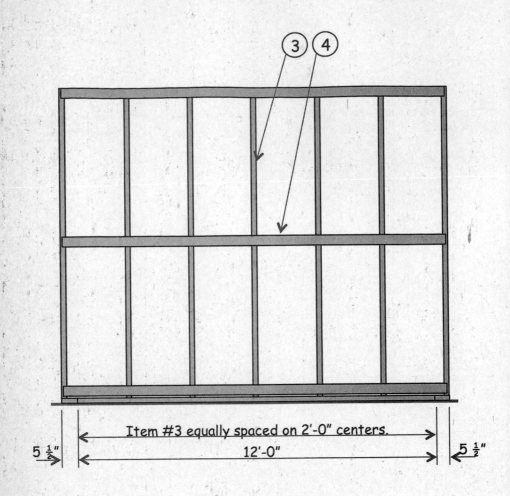

Side View

Step 2: Attach Wire Mesh:

Attach wire mesh item
#7 to frame using heavy
duty staples item #8.
Cover frame with UV
stabilized polyethylene
item #9.

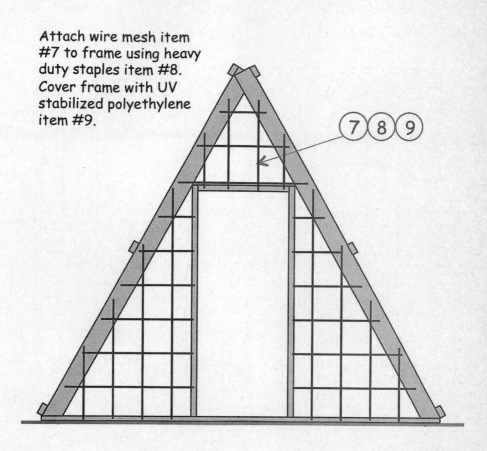

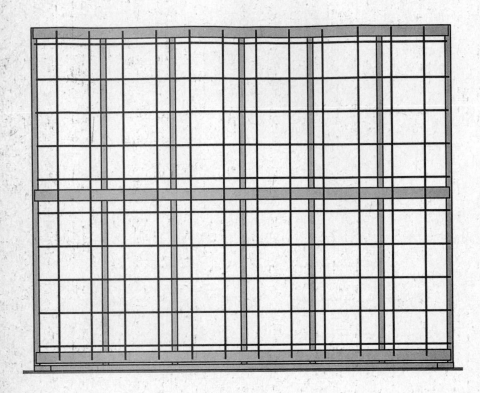

Side View

Step 3: Install Door:

Install pre-hung door
item #10 using
hardware supplied with
door.

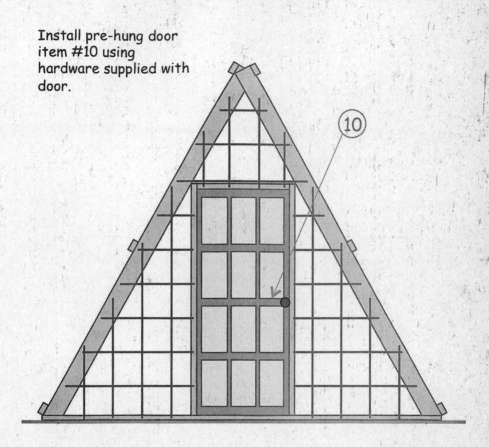

Wooden Arched "A" Frame Greenhouse 8x10

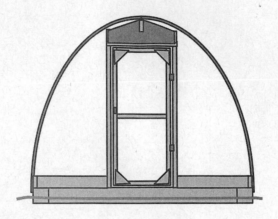

Bill of Material

Item	Qty	Description	Item	Qty	Description
1	2	8x8 x 8'-0"	11	2	1/2" plywood x 1'-2" x 3'-1"
2	2	8x8 x 10'-0"	12	1 box	5d nails x 1 3/4" LG
3	4	2x6 x 3'-0"	13	4	1/2" plywood x 6" x 18'-0" (splice)
4	2	2x6 x 10'-0"	14	4	2x4 x 5'-11"
5	1 box	10d nails x 3" LG	15	6	2x4 x 2'-6"
6	4	2x4 x 6'-0"	16	8	1/2" plywood x 6" x 6"
7	2	2x4 x 3'-1"	17	4	2" hinges
8	2	2x6 x 10'-0"	18	2	door handles
9	1	2x6 x 10'-0"	19	2	door latch
10	2	2x6 x 8 1/2"	20	1 roll	UV resistant polyethylene

Step 1: Assemble Base:

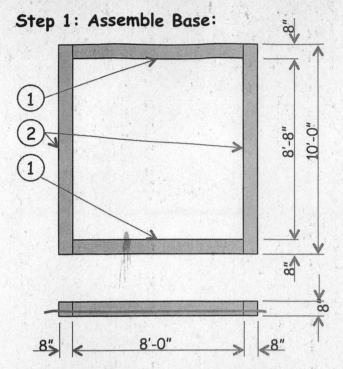

Place base timber
item #1 & #2 as
shown in flat area.

Step 2: Assemble Lower Frame:

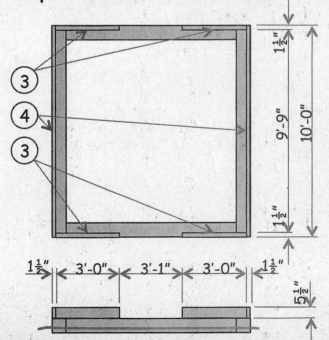

Assemble lower
support frame item
#3 & item #4 using
10d nails item #5.

Step 3: Assemble End Frames:

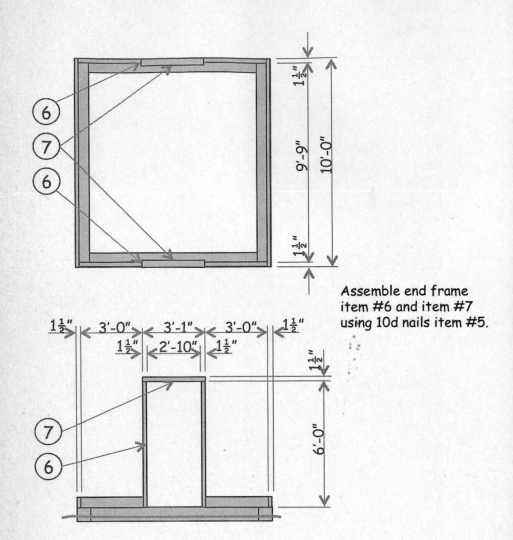

Assemble end frame item #6 and item #7 using 10d nails item #5.

Step 4: Install Roof Supports:

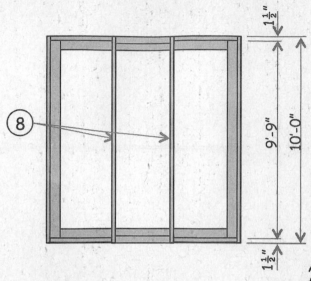

Assemble roof supports frame item #8 using 10d nails item #5.

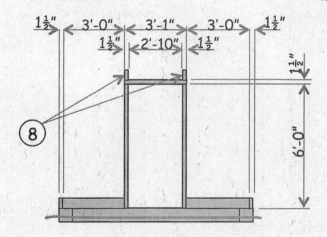

Step 5: Install Roof Supports:

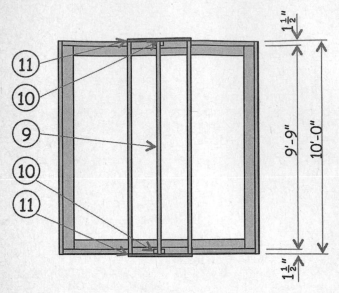

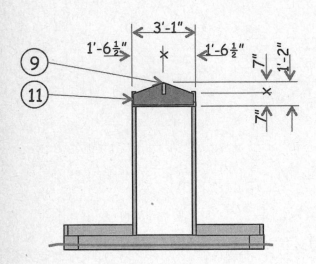

Assemble upper roof support item #9, #10, and #11 using 5d nails item #12.

Step 6: Install Roof Supports:

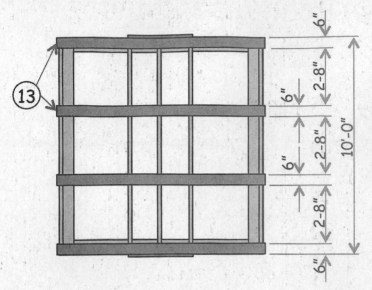

Assemble
support arch
item #13
using 5d nails
item #12.

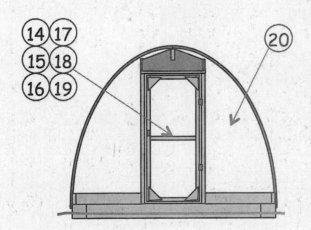

Assemble end doors.
Cover structure with UV
resistant polyethylene.

Pole Barn Greenhouse Plans 22 FT x 32 FT

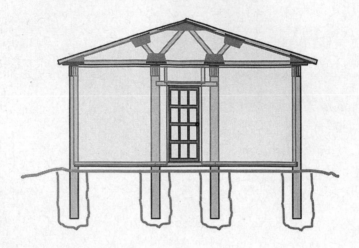

Bill of Material

Item	Qty	Description	Item	Qty	Description
1	20	6x6 x 12'-0"	12	34	2x4 x 5'-9"
2	20bags	readymix concrete	13	34	2x4 x 12'-0"
3	1 box	10d nails x 3" LG	14	17	1/2" plywood x 1'-0" x 3'-0"
4	37	2x10 x 16'-0"	15	68	1/2" plywood x 6" x 1'-0"
5	40	2x2x 1/4" angle x 6" LG	16	1 box	5d nails x 1 3/4" LG
6	12	2x4 x 8'-0"	17	50	clear fiberglass panels
7	1 box	3/8" lag bolts x 3" LG	18	1 box	#8 screws x 1 1/2" LG
8	4	2x4 x 6'-0"	19	2	prehung door
9	4	1/2" plywood x 6" x 12"	20	4	2" door hinges
10	34	2x4 x 13'-0"	21	2	door handle
11	34	2x4 x 3'-0"	22	2	door latch

Step 1: Install Posts:

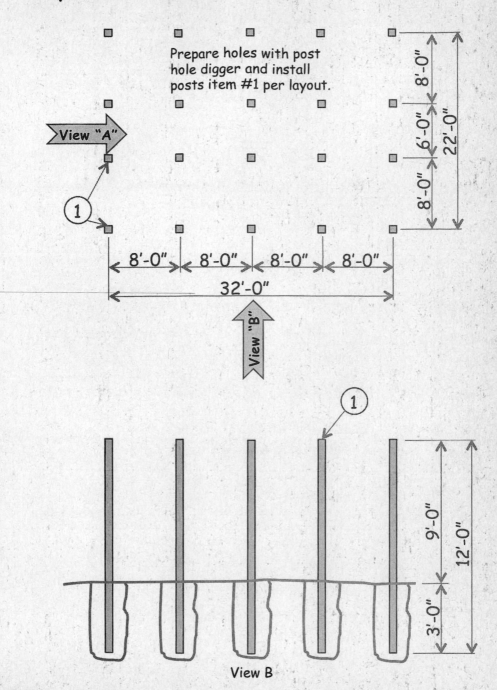

Prepare holes with post hole digger and install posts item #1 per layout.

View B

Step 3: Install Truss Supports:

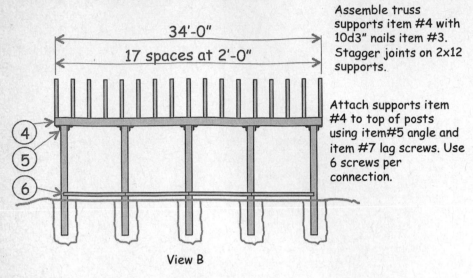

Assemble truss supports item #4 with 10d3" nails item #3. Stagger joints on 2x12 supports.

Attach supports item #4 to top of posts using item#5 angle and item #7 lag screws. Use 6 screws per connection.

View B

Step 4: Install Roof Truss Assemblies:

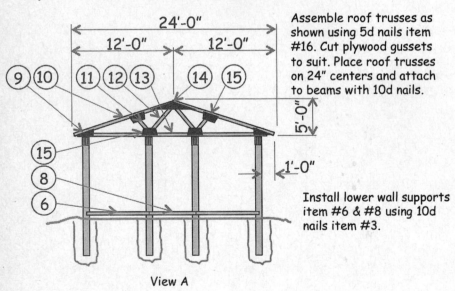

Assemble roof trusses as shown using 5d nails item #16. Cut plywood gussets to suit. Place roof trusses on 24" centers and attach to beams with 10d nails.

Install lower wall supports item #6 & #8 using 10d nails item #3.

View A

Step 5: Final Assembly:

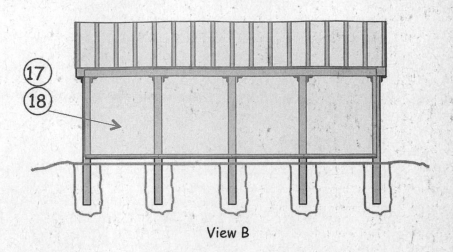

⑰
⑱

View B

Install pre-hung door
assembly. Cover structure
with clear fiberglass panels.

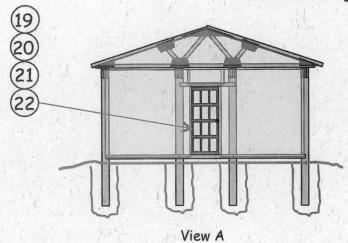

⑲
⑳
㉑
㉒

View A

Greenhouse Plan
12 FT x 32 FT

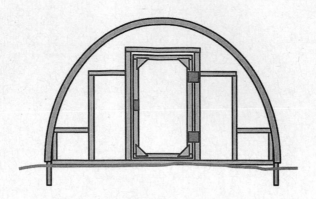

Bill of Material

Item	Qty	Description	Item	Qty	Description
1	34	1/2" rebar x 2'-6"	11	4	2x4 x 3'-8 1/2"
2	4	2x6 x 16'-0"	12	4	2x4 x 5'-3"
3	2	2x6 x 12'-0"	13	1 box	#8 screws x 1 1/2" LG
4	1 box	10d nails x 3" LG	14	34	PVC pipe x 10'-0"
5	4	2x4 x 1'-6"	15	34	PVC pipe couplings
6	8	2x4 x 2'-0"	16	1 box	wire ties
7	4	2x4 x 4'-7"	17	4	2" hinges
8	4	2x4 x 5'-7"	18	2	door handle
9	2	2x4 x 4'-0"	19	2	door latch
10	8	1/2" plywood x 6" x 6"	20	1 roll	UV resistant polyethylene

Step 1: Install Support Stakes:

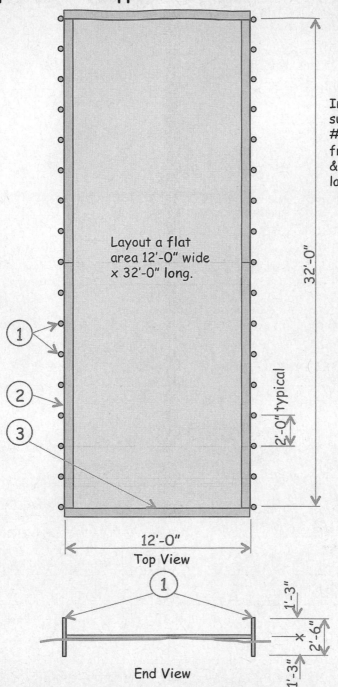

Install hoop supports item #1 and base frame item #2 & item #3 per layout.

Layout a flat area 12'-0" wide x 32'-0" long.

32'-0"

2'-0" typical

12'-0"

Top View

End View

1'-3"

2'-6"

1'-3"

Step 2: Assemble End Frames:

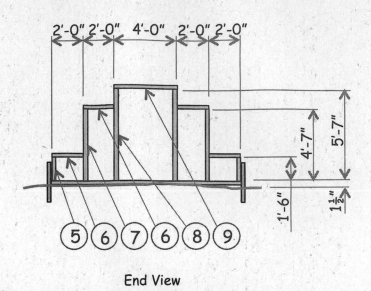

End View

Step 3: Assemble Doors:

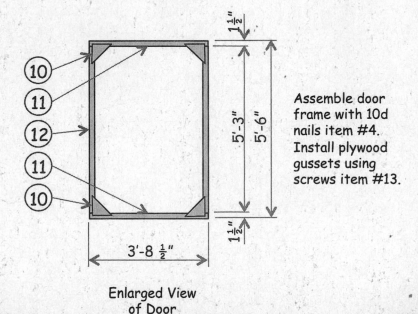

Assemble door frame with 10d nails item #4. Install plywood gussets using screws item #13.

Enlarged View
of Door

Step 4: Final Assembly:

Attach PVC pipe item #14 end to end using PVC couplings item #15. Place PVC pipe over rebar and secure with wire ties item #16. Install door assembly using hinges item #17, latch item #18, and handle item #19. Cover entire structure with UV resistant polyethylene item #20.

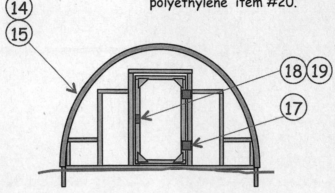

End View

Greenhouse Plan
Mini PVC Frame
6 FT x 6 Ft

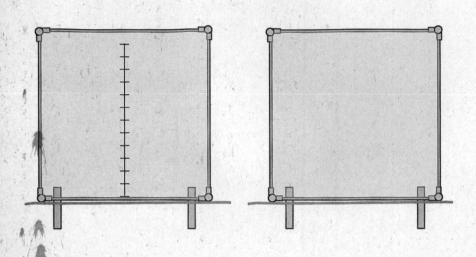

Bill of Material

Item	Qty	Description	Item	Qty	Description
1	8	1" PVC corner fittings	5	4	metal pipe straps
2	12	1" PVC pipe x 6'-0"	6	1 box	#8 wood screws x 2" LG
3	1 can	PVC adhesive	7	1 roll	UV resistant polyethylene
4	4	2x2 wood stakes x 18" LG	8	1	6 FT long zipper

Step 1: Assemble Base Frame:

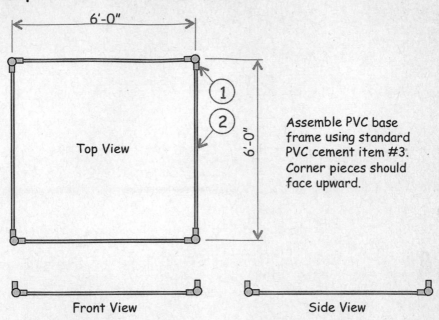

6'-0"

6'-0"

Top View

Front View

Side View

Assemble PVC base frame using standard PVC cement item #3. Corner pieces should face upward.

Step 2: Install Vertical Legs:

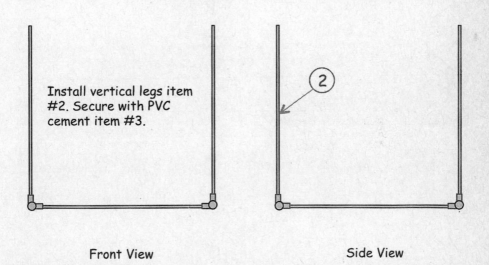

Install vertical legs item #2. Secure with PVC cement item #3.

Front View

Side View

Step 3: Assemble Top Frame:

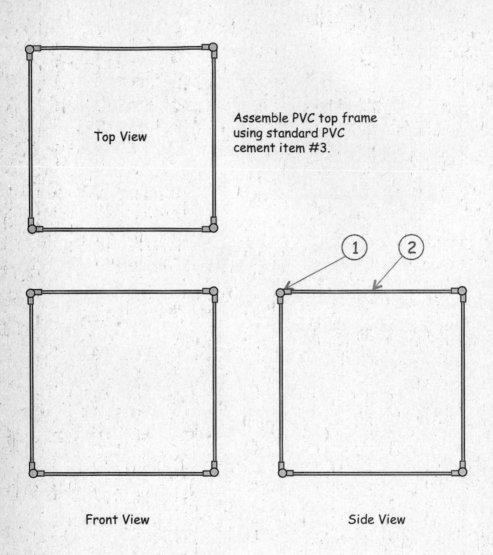

Top View

Assemble PVC top frame
using standard PVC
cement item #3.

Front View

Side View

Step 4: Final Assembly:

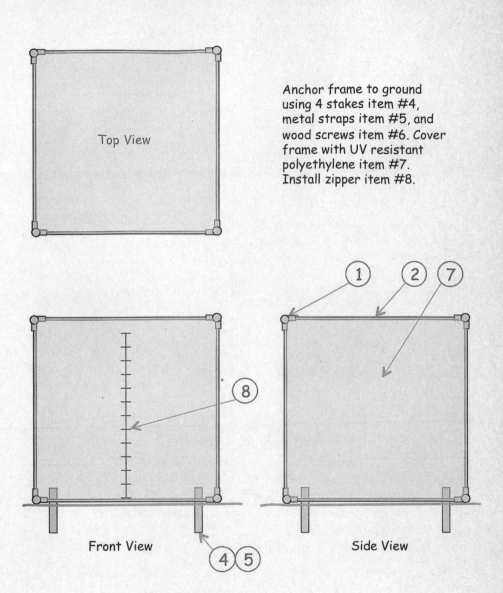

Anchor frame to ground
using 4 stakes item #4,
metal straps item #5, and
wood screws item #6. Cover
frame with UV resistant
polyethylene item #7.
Install zipper item #8.

Top View

Front View

Side View

Greenhouse Plan
24x48
Potting Bench

Bill of Material

Item	Qty	Description		Item	Qty	Description
1	2	2x4 x 4'-0"		4	2	2x4 x 2'-10"
2	4	2x4 x 2'-0"		5	1 box	#8 wood screws x 2 1/2" lg
3	4	2x4 x 4'-0"		6	7	2x8 x 4'-3"

Step 1: Assemble Frame:

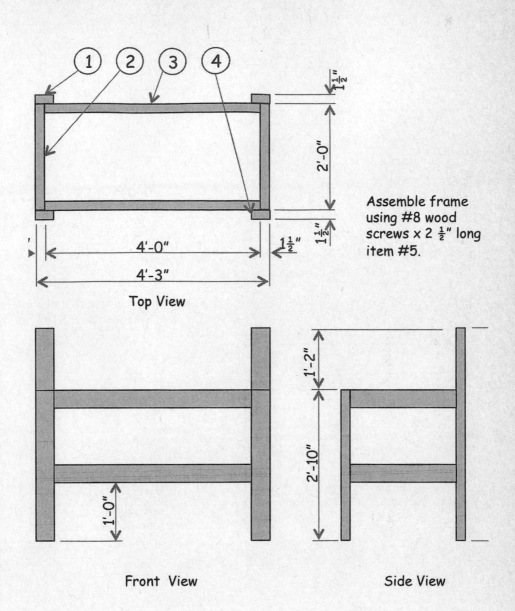

Assemble frame using #8 wood screws x 2 ½" long item #5.

Top View

Front View

Side View

Step 2: Install Shelving:

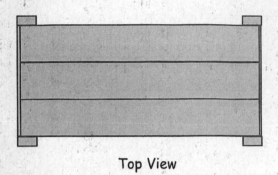

Top View

Assemble shelving item #6 using #8 wood screws x 2 ½" long item #5.

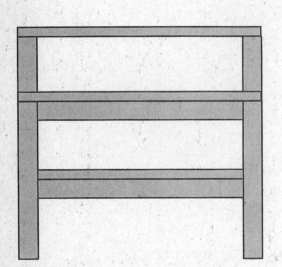

Front View

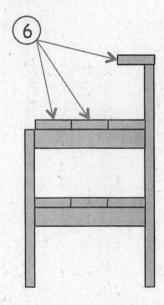

Side View

Green House Planting Bench Plan 30" x 72"

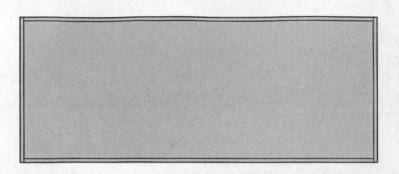

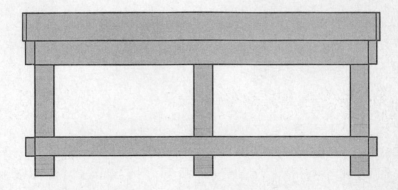

Bill of Material

Item	Qty	Description	Item	Qty	Description
1	2	2x6 x 6'-0"	6	2	2x4 x 2'-6"
2	6	4x4 x 2'-6"	7	1	1/2" plywood x 2'-6" x 6'-3"
3	2	2x6 x 2'-6"	8	2	1/2" plywood x 6" x 2'-7"
4	1 box	#8 screws x 3" LG	9	2	1/2" plywood x 6" x 6'-3"
5	2	2x4 x 6'-0"	10	1 box	#8 screws x 2" LG

Step 1: Assemble Base Frame:

Assemble base frame using #8 screws x 3" long item #4.

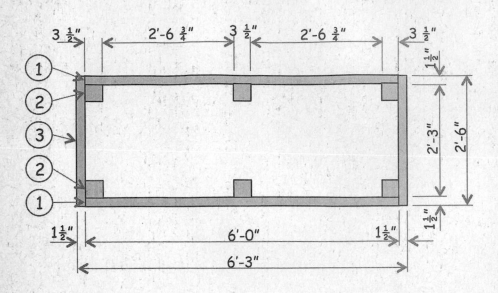

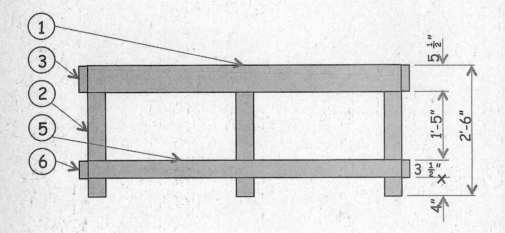

Step 2: Assemble Planting:

Assemble planting box using #8 screws x 2" long, item #10.

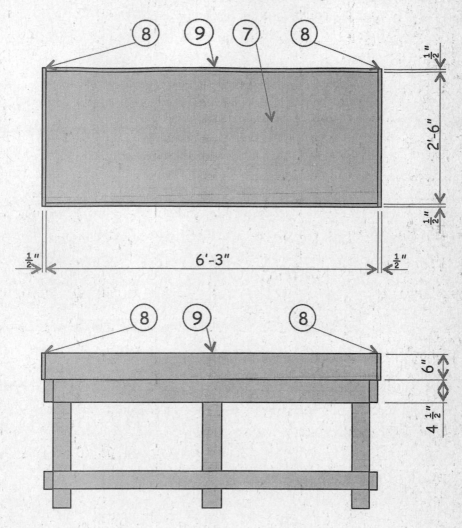

Conclusion

You may have thought that building a greenhouse only involved putting the structure together, but the truth is building a greenhouse is as much about what is inside the greenhouse, as the greenhouse itself. Whether you decide to build your own greenhouse from scratch, hire a professional, or build it from a pre-fabricated kit, it helps if you have a basic knowledge of the process to guide you through each stage of construction. Understanding these basics will help you to possess more than a general idea of how it is done. It will help if you become familiar with the important steps along the way to help ensure that your greenhouse functions properly after the construction has been completed.

As with many craft projects, some people are more "do-it-yourself" types than others. Performing carpentry and electrical work or building plumbing systems are not especially challenging for them. Others prefer to call the professionals for assistance or to just start with a simpler project. One of the nice things about greenhouses is that there is a level of expertise for everyone. That means a gardener can do what he or she loves best: planting and sowing.

Hopefully, this book made it easier for you to decide whether you should purchase and build a greenhouse kit or construct your own from the ground up. Regardless of your decision, you are now familiar with the step-by-step instructions for building a greenhouse and the basic principles of how a well-constructed greenhouse works inside and out.

Glossary

Auger – A powered tool used to bore holes in the ground.

Batten tape – A kind of tape used for securing greenhouse film to wood baseboards.

Batter board – A horizontal board that is attached to posts located at the corners of the building site that shows the accurate layout of a foundation.

Batts – Any insulation that is cut into sheets, rather than in loose pieces.

Bench – A portable work surface that is used to hold pots and trays of seeds.

Blanket – A type of insulation made of batts or rolls of fiberglass.

BTU or Btu – A traditional calculation that stands for British thermal units.

Caulking – The process of sealing joints or seams for piping in a structure.

Caulking gun – A pistol-like device loaded with a caulking cartridge to seal objects.

Cedar – A high-quality wood characterized by a pleasant, fresh smell. It is red in color.

Chisel – A tool with a cutting edge, used for carving wood, stone, or metal.

Combustion – An exothermic chemical reaction caused by the combination of fuel, an oxidant, and heat.

Concrete – A building material made of a mixture of cement, gravel, sand, and water.

Condensation – The change in matter from a gas into liquid droplets.

Conduction – The transfer of heat, or thermal energy, between molecules.

Convection – The heat transfer within an environment.

Even-span (traditional-span) – Common type of greenhouse that can be attached to a house or it can be a freestanding structure.

Fasteners – A hardware device that secures two or more objects together.

Framing square – A tool with a long and short arm that meet to make a 90-degree angle.

Gable – A triangular part of the wall between the edges of a sloped roof. This type of greenhouse typically has vertical sidewalls and an even-span roof with plenty of headroom along the center of the structure.

Hammer – A tool used to deliver impact to an object, usually a nail.

Heliotropic – The turning or bending of a plant toward or away from sunlight.

Junction box – A container, usually in the shape of a square, which houses and conceals electrical connections.

Knee wall – An architecturally short wall, usually less than 3 feet in height.

Level – A glass tube that contains liquid with an air bubble inside it, to determine if the site is basically even.

Louver – A window or blind with adjustable, horizontal slats that are set at an angle to let in light and air, but to repel rain, snow, and direct sunlight.

Mallets – A hammer-like tool usually made of wood, with a skinny handle and an oversized head.

Microclimate – A small ecosystem created in a localized space where the climate inside is different than the surrounding climate.

Orangery – An unheated structure with a solid roof and glass walls.

Photosynthesis – The process of using the sun as a way of converting carbon dioxide into sugars.

Plane – A flat, two-dimensional surface with a point, a line, and a space.

Plumb – A vertical surface, such as a wall or door, is perpendicular to a horizontal surface.

PVC (polyvinyl chloride) – A tough, synthetic resin that is made with polymerizing vinyl chloride.

Quonset – A lightweight structure usually made of galvanized steel with a semi-circular section.

Radiation – Any process in which energy travels through a medium or space and is absorbed by another body.

Rafters – Support boards laid out across the plate at the top of a wall.

Roof eave – The edge of a roof that projects past the side of the building.

Saw – A construction tool with a handle and a hard blade with small abrasive edges used to cut through a material.

Screwdriver – A tool used for rotating screws into a surface.

Studs – Vertical posts that support a wall.

Sunroom – A structure built onto the side of a house with windows to let in sunlight.

Tape measure – The flexible form of a ruler.

Turnbuckles – A device used for adjusting the tension on ropes and cables.

Ventilation – The process of changing or replacing air in an indoor area.

Washer – A thin plate with a hole in the middle that distributes the load of a threaded fastener.

Water line – A pipe that moves water from one location to another.

Wire – A single string of metal used to bind mechanical loads and carry electrical signals.

Wrenches – A tool used to provide grip and apply torque to turn an object, such as a nut.

Author Biography

Craig Baird is a writer based out of rural Canada, where he lives on a ranch with his wife and dogs. He has published several books and short stories, as well as written for magazines and newspapers across Canada. When he is not writing, he spends his time traveling the country and hiking in the outdoors with his wife, Layla.

Index